"If you're going to read only one book about narcissism this is it. It's not only eminently accessible for the lay audience and professional alike, but it also offers a profound analysis of the nature of narcissism and common misunderstandings about it. Add to this Dr. Malkin's penetrating insights, his superb ability to tell a good story, and his courage in disclosing elements of his own story, and you wind up with this remarkable book."

—Joseph Shay, PhD, lecturer at Harvard Medical School, coauthor of *Psychodynamic Group Psychotherapy*, and coeditor of *Odysseys in Psychotherapy* and *Complex Dilemmas in Group Therapy*

"A fresh approach to the way we regard one of psychology's most complex conditions. In a book that's persuasive, insightful, and never dry, Dr. Malkin offers the right mix of analysis and advice, and presents compelling, groundbreaking evidence that narcissism is necessary—in the right doses, of course."   —Peggy Drexler, PhD, assistant professor at Weill Cornell Medical College and author of *Raising Boys Without Men* and *Our Fathers, Ourselves*

"This is an enthralling book. It takes the clichés of narcissism and unpacks them to help us understand and accept our human need to feel special while also coping with the dangers of self-absorption. It will become a classic."

—Dr. Sue Johnson, author of *Love Sense: The Revolutionary New Science of Romantic Relationships*

"[A] book that will have readers rethinking themselves and, paradoxically, those around them."   —*Publishers Weekly*

"This is a true gem on the subject of narcissism."

—*Library Journal*

"[Dr. Malkin's] reassuring tone and plethora of case histories offer considered advice and generous encouragement."

—*Kirkus Reviews*

"A gripping and sometimes terrifying book that will make you look anew at your spouse, your parents, your children, your friends, your enemies, your fellow workers, and—perhaps most pertinently—your reflection in the mirror."

—*The Daily Mail* (London), "Book of the Week"

"Dr. Craig Malkin offers a surprising, accessible analysis of narcissism—and explains why a healthy dash of narcissism can be a good thing." —Gretchen Rubin, bestselling author of *Better Than Before* and *The Happiness Project*

"In *Rethinking Narcissism*, Dr. Malkin reveals the surprising good news about narcissism, exploring the complexities of narcissistic traits and deflating popular myths. Most importantly, he shows us how to develop a healthy sense of narcissism and how to manage relationships with narcissistic partners, friends, colleagues, and family."

—Dr. Drew Pinsky, author of *The Mirror Effect*

"Certainly one of the best books I've read this year. Don't be fooled by the title . . . this book is for anyone trying to better understand themselves and other people."

—Todd Kashdan, PhD, author of *The Upside of Your Dark Side*

"[A] fascinating book." —*The Independent* (London)

"Among all the books that have been published on the topic in the past ten years, *Rethinking Narcissism* . . . stands out as a definite must-read." —*Psych Central*

"Thank you, Dr. Malkin, for saying what needed to be said and clearing things up for me. For all of us." —*BookTrib*

"If you're to buy just one book on narcissism, this is the one to purchase." —Leon Seltzer, author of *Paradoxical Strategies in Psychotherapy*

"Malkin, a therapist and psychology instructor at Harvard Medical School, takes a more inspirational attitude. . . . " —*New York Times Book Review*

"[G]ives us all a coherent way of talking about a much-discussed but often oversimplified and overdramatized subject in these 'look at me' times." —Peg Streep, bestselling author of *Mean Mothers: Overcoming the Legacy of Hurt*

"*Rethinking Narcissism* brings much-needed compassion and clarity to one of the most vexing problems in mental health without ever resorting to false hope or naïveté. In that way, the book itself is special." —Tom Wootton, *Huffington Post* blogger and author of *The Bipolar Advantage*

"The book that protects you from narcissists. . . . Is there someone in your life who's hurting you and you just don't know it? In this Harvard researcher's illuminating, reads-like-a-novel book, he reveals how to identify and repair your relationships to live with more fulfillment."

—Oprah Book Club 2.0

"Is there a narcissist in your life? Chances are, the answer's yes—here's how to spot them."   —*Red* magazine

"Narcissists. They're everywhere. . . . Not according to Dr. Craig Malkin, whose new book suggests we've got it all wrong."   —*Sunday Times Magazine* (London)

RETHINKING
NARCISSISM

THE SECRET TO RECOGNIZING
AND COPING WITH NARCISSISTS

DR. CRAIG MALKIN

HARPER  PERENNIAL

NEW YORK • LONDON • TORONTO • SYDNEY • NEW DELHI • AUCKLAND

For Julie Malkin

HARPER ● PERENNIAL

A hardcover edition of this book was published in 2015
by HarperCollins Publishers.

HarperCollins books may be purchased for educational,
business, or sales promotional use. For information
please e-mail the Special Markets Department at
SPsales@harpercollins.com.

FIRST HARPER PERENNIAL EDITION PUBLISHED 2016.

Designed by Jo Anne Metsch

Illustration on page 1: *Echo and Narcissus*,
by John William Waterhouse, 1903.

The Library of Congress has catalogued the
hardcover edition as follows:

Malkin, Craig.
Rethinking narcissism : the secret to recognizing
and coping with narcissists / Dr. Craig Malkin.
Pages cm
ISBN 978-0-06-234810-4
1. Narcissism. I. Title.
BF575.N35M353 2015
155.2'32—dc23
2015006017

ISBN 978-0-06-234811-1 (pbk.)

23 24 25 26 27 LBC 20 19 18 17 16

CONTENTS

PART III: RECOGNIZING AND COPING WITH UNHEALTHY NARCISSISM

PART IV: PROMOTING HEALTHY NARCISSISM

INTRODUCTION

My mother was the most wonderful and infuriating person I've ever known: she was a narcissist.

I wasn't aware of it for the longest time, not until I was in college and immersed in an introductory psychology text. There, printed in bright bold letters just below a picture of the Greek youth Narcissus staring at his reflection in a pool of water, was the word *narcissism*. When I read the accompanying description, I remember feeling relieved and horrified all at once. The term perfectly captured the paradox of my mother.

She was the incandescent figure of my childhood, irrepressibly outgoing, infectiously funny, and wonderfully caring. The world seemed to revolve around her. A striking nearly six-foot-tall blonde, with a thick English accent from her upbringing in Great Britain, she seemed to make connections everywhere she went—the grocery store, the coffee shop, the hair salon. She was devoted to friends, buoying them through illness and hardships, and dedicated to improving her community, whether the project was cleaning up a playground or organizing a bake sale.

And as wife to my father and mother to me and my brother, she was always there, generous with her love and counsel.

But her glow gradually dimmed as I, and she, grew older. She seemed to become more self-involved. She bragged about her accomplishments as a young ballet dancer, sometimes making the point by demonstrating—awkwardly—a split or plié. She name-dropped, boasting of brushes with celebrities (though I could never tell if the encounters were real or imagined). She grew obsessed with her looks, frantically charting wrinkles and chasing spots around her body and starving herself to stay thin. She interrupted people when they spoke, even when they were in the midst of sharing their pain and anxiety. Once, when I tried to tell her of my anguish over a romantic breakup, she dreamily muttered, "I never had any trouble finding dates." I was stunned by the non sequitur.

What had happened to my mother? College gave me the word *narcissism*. But I really didn't understand what it meant. I had so many questions. Had she always been a narcissist and I hadn't recognized it? Was she suddenly pushed to it by circumstance, namely getting older? Could I do anything to get back the loving, unselfish woman I remembered from my childhood?

I devoted myself to finding answers. In the library, I pored over books and articles from Freud onward. As a psychologist in training, I interned with one of the foremost experts on narcissism. I took a postdoctoral fellowship focused on helping personality-disordered clients, hoping to better understand narcissistic personality disorder (NPD), the most extreme form of narcissism. But even though I learned a great deal during those years, my understanding still felt incomplete. Then one day, I saw something that changed my thinking about narcissism—in my mother, in my clients, and in myself—forever.

My father had recently died and my wife, Jennifer, and I had undertaken the painful process of moving my mother from a large house far away into a small apartment close to us. The cramped space she found herself in now pushed her over the edge. "Lovely place you've found for me," she grumbled sarcastically.

She stayed in a nearby hotel that first night, rolling up in a taxi the next afternoon to meet us at the apartment. We resumed unpacking, mostly in silence and mostly without her help. Before long, my mother disappeared in a taxi again, this time to drop exorbitant sums on "decorations."

It went on that way for a week—my mother staying nights in a hotel, shopping by day—until late one evening, she announced, with an exaggerated sigh, "I need to get comfortable!" She disappeared into the bedroom where we heard her rustling through boxes. Moments later, she reappeared wearing four-inch stilettos—Manolo Blahniks, she proudly informed us. "There," she said, sighing, "I can relax now. At least my shoes are better than this place." The shoes, apparently, made her feel special.

That's when it hit me. My mother used feeling special as a crutch—something to prop herself up when she felt scared or sad or lonely. Instead of turning to me, my brother, Jennifer—or anyone—to talk about how frightened she was about living alone, she relied on feeling better than other people. (In her Manolos, she literally was above most people.) It hadn't been so necessary to make herself feel special when she was younger—others did the job with their attention and compliments. But as she aged and her looks—the source of much of her confidence—faded, she grew to believe that she had very little to offer and she withdrew from social and civic life. She had to find another way to stand out and prove to herself that she was special.

Thinking of narcissism this way—as a habit people use to comfort themselves—showed me a much clearer, simpler path to coping with my mother. I could see what made her narcissism rise and fall. I could see how and why it became destructive. I could even see how to help her to set it aside and talk honestly about her pain.

My search to understand my mother led me to another epiphany as well: narcissism isn't all bad. In fact, some narcissism is good—even vital—for us to lead happy, fulfilled, and productive lives. Feeling special, I've discovered, can make us better lovers and partners, courageous leaders, and intrepid explorers. It can make us more creative, and it might even help us live longer.

Numerous studies confirmed much of what I'd seen growing up. The traits I so admired in my mother when she was young—her warmth, optimism, and activism—were fueled in great part by her narcissism. Her sense that she was special gave her conviction, confidence, and courage. It allowed her to believe that she had wisdom to effect change in the world, the ability to pull off just about anything she set her mind to, and the nerve to go ahead and try. Narcissism was her launching pad. It made her an engaged parent and energetic community leader. And it made her believe not just in herself but in others as well—and they felt that assurance.

When I was seven, I remember her talking to a despairing shop owner who was very close to shutting his doors. "We need you," she said, beaming. "I need you. Where else would I get such perfectly delicious food and brilliant conversation?" Her lips formed an exaggerated pout. "That's it!" she said, stamping her foot. "You cahhhhn't leave—I won't have it!" Munching on cookies, I watched the man's face go from crestfallen to triumphant. Such was the power of my mother; she felt special and she made others feel special, too. The man's store stayed open well into my college years.

That feeling special can be good as well as bad is just one of the startling findings I unearthed while exploring the mystery of narcissism. In the following pages you'll discover many other truths that challenge accepted wisdom. In reaching my conclusions, I've drawn from a wealth of research, much of it conducted during the past few years. I've also drawn on my experience as a clinician working with individuals and couples to provide vivid examples of narcissism at its worst, its best, and in all its subtleties. (All the examples are composites of people I've counseled; identifying information has been changed to protect people's privacy.)

My goal in writing this book is to help you not only understand and cope with the people around you—those you live and work with—but also to better understand yourself. My explorations certainly did that for me.

Like many children of narcissists, while growing up and through my teen years, I didn't allow myself to feel special at all. I was terrified of even trying. I shrank from compliments or dismissed them. No matter what I accomplished, it wasn't good enough.

Later as a young adult struggling to find my voice, I swung in the opposite direction, dominating conversations with one too many jokes or tall tales, all in an effort to prove I had something interesting to say. What I eventually realized is that neither stance—constant self-doubt or continuous bravado—made for a very fulfilling life; they both left me feeling lonely and misunderstood.

Luckily, I've been able to change and find a rewarding balance. And I've helped many others do the same. As a therapist, I am a firm believer that growth is possible, for everyone, whether we harbor too little narcissism or too much. And happily, the evidence, as you'll see, supports that conclusion.

Years after I started researching this book, in the midst of a particularly blistering summer, my mother passed away. My brother and I were at her side. By that time, I had come to see her narcissism in a different, more nuanced light. Without that new perspective, I'm certain I wouldn't have been able to say goodbye to her with love in my heart.

My aim in sharing the insights you're about to discover is to bring the same clarity and hope to your life that I found in my own.

May this book help you overcome the bad—and embrace the good—about feeling special.

THE MYTH OF NARCISSUS

Long ago in Ancient Greece there lived a boy, Narcissus, the son of the river god Cephissus and the fountain nymph Liriope. His divine origin had blessed him with equally divine looks. With wavy locks tumbling over his forehead and a body sculpted by years of climbing trees and scrambling over rocks hunting for deer and birds, Narcissus quickly amassed an army of admirers.

People everywhere—young and old, men and women—fell for him almost instantly. Soon his reputation reached beyond the human world. Anytime he wandered through the thick forests or along the rippling rivers near his home, Narcissus inevitably drew a crowd of tree or water nymphs eager to catch a glimpse of him.

Narcissus became accustomed to this admiration, but never offered a welcoming response. As legendary as his beauty might have been, he soon became equally well known for his indifference. One by one, potential lovers approached him and, one by one, he turned them away. He seemed to think himself above kindness or love, above the ordinary world of humans, above everyone, really—even the gods.

One day the mountain nymph Echo joined the ranks of his unrequited lovers. As the sun broke through the trees of the forest she caught a glimpse of Narcissus strolling through the woods on his daily hunt. Her heart burned. Unable to look away, she began to follow him, discreetly at first, peering quietly through the branches and leaves. Then, overcome by passion, she grew bolder, trampling noisily in his path. Soon, he sensed he was being followed.

"Who's there?" he called.

Echo tried to answer, but she had no voice of her own—the result of an ancient curse by the goddess Hera (Echo had distracted her with incessant chatter one too many times). She tried to call out, but could only repeat his words.

"Who's there?" she replied sadly.

"Come out now!" he demanded.

"Out now," she answered, tearfully.

Growing angry, perhaps feeling mocked, Narcissus shouted. "Show yourself!"

"Yourself!" cried Echo, leaping out from behind the trees. She reached out, throwing her arms around his neck.

But Narcissus's heart remained cold. "Get away!" he barked. Then, as he fled, he yelled cruelly over his shoulder, "I'll die before I love you!"

"Love you!" Echo called, sobbing. Humiliated and heartbroken, she disappeared into the thickest part of the woods. She *refused to move, refused even to drink or eat, and her body slowly withered away, until only her voice remained.*

Meanwhile, the gods grew tired of the wreckage Narcissus had been leaving in his wake. One man, Ameinias, had become so distraught when Narcissus spurned his advances that he drew a sword and ran himself through. But before he did, he whispered a prayer to the goddess of vengeance, Nemesis. She quickly answered with a curse befitting the cruelty she'd witnessed. Narcissus himself was to know the pain of unrequited love.

One afternoon soon after, while strolling through his beloved woods, he came upon a cool, clear spring, so eerily still that it looked like a mirror. Thirsty from the walk, he bent down to drink, and when he did, he caught a glimpse of a beautiful face. He was so clouded by Nemesis's curse he didn't realize he was staring at himself. His heart hammered in his chest. He'd never known a feeling like this before, the depth of longing, the sheer joy of being in a person's presence. Maybe this is love, he thought.

"Come join me!" he cried.

Silence.

"Why won't you answer me!" he bellowed, gazing at his reflection. "Don't you want me, too?"

He bent down to kiss the water and the face, briefly, seemed to fade from view.

"Come back!" He tried to approach the man again, to touch him, to feel his embrace. But each time he did, the face seemed to retreat, disappearing into the still waters of the spring.

Hours went by, then days, until at last, Narcissus stood up and dusted himself off. He finally knew what to do.

"I'll come to you!" he called out into the water. "That way we can be together!"

With that, he dove into the pool, plunging down into the

darkness, deeper and deeper, until he disappeared from sight, never to surface again.

Moments later, at the edge of the pool, a fantastic flower sprang up, a nimbus of white petals ringing a bright yellow trumpet. It leaned over the pool, forever gazing into the waters beneath it.

PART I

1

RETHINKING NARCISSISM

OLD ASSUMPTIONS, NEW IDEAS

> The silent killer of all great men and women of achievement—particularly men, I don't know why, maybe it's the testosterone—I think it's narcissism. Even more than hubris. And for women, too. Narcissism is the killer.
> —BEN AFFLECK

Narcissism. The word has soared to such dizzying heights of fame that Narcissus himself would flush with pride. Scan a newspaper or magazine, watch the nightly news or daily talk shows, eavesdrop on commuters on their cellphones, gossip with your next-door neighbor, and the word pops up again and again. Everyone's using it: average citizens, actors, social critics, therapists, a US Supreme Court justice, even the pope. Add in that we're allegedly in the midst of a "narcissism epidemic," and it's easy to see why the term has become ubiquitous. Nothing gets people talking like a disease on the rise, especially if, as Ben Affleck seems to worry, the condition is terminal.

But what does *narcissism mean* exactly? For a word that gets hurled about with such frequency and fear, its definition seems alarmingly vague. Colloquially, it's become little more than a popular insult, referring to an excessive sense of self—self-admiration, self-centeredness, selfishness, and self-importance. The press is apt to slap the description on any celebrity or politician whose publicity stunts or selfie habits have spiraled out of control.

But is that all narcissism is? Vanity? Attention-seeking? In psychological circles the meaning is no less confusing. Narcissism can either be an obnoxious yet common personality trait or a rare and dangerous mental health disorder. Take your pick. But do it soon, because there's a strong sentiment among mental health researchers that it shouldn't be considered an illness at all.

As slippery and amorphous as all these views seem to be, they all share a single assumption: narcissism is *wholly* destructive.

Too bad it's wrong.

Narcissism can be harmful, true, and the Web is rife with articles and blogs from people who've suffered at the hands of extremely narcissistic lovers, spouses, parents, siblings, friends, and colleagues. Their stories are as heartbreaking as they are frightening. But that's just a small part of narcissism, not the whole picture. And until all the pieces are in place, we have little hope of understanding how and why narcissism becomes destructive, let alone protecting ourselves when it does.

Today, in contrast, a surprising new view has begun to emerge, one that points to all the ways narcissism seems to help us, too. It even offers some hope for change when our loved ones, just like Narcissus, are in danger of disappearing into themselves forever.

Narcissism is more than a stubborn character flaw or a severe mental illness or a rapidly spreading cultural disease, transmitted by social media. It makes no more sense to *assume* it's a problem than it would if we were speaking of heart rate, body temperature, or blood pressure. Because what it is, in fact, is a normal, pervasive human tendency: *the drive to feel special*.

Indeed, for the past twenty-five years, psychologists have compiled massive amounts of evidence that most people seem convinced they're better than almost everyone else on the planet. This wealth of research can only lead us to one inevitable conclusion: the desire to feel special isn't a state of mind reserved for arrogant jerks or sociopaths.

Consider, for example, the findings from a research tool called the "How I See Myself Scale," a widely used questionnaire devised to measure "self-enhancement" (an unrealistically positive self-image). People who fill out the scale are asked to rank themselves on various traits, including warmth, humor, insecurity, and aggressiveness ("Do you think you're average or in the top 25 percent, 15 percent, or 10 percent?"). In study after study in country after country, the vast majority of participants report having more admirable qualities and fewer repugnant ones than most of their peers. After reviewing decades of findings, University of Washington psychologist Jonathan Brown has concluded, "Instead of viewing themselves as average and common, most people think of themselves as *exceptional* [emphasis added] and unique." This pervasive phenomenon has been dubbed "the better than average effect."

Lest you fear that these results are evidence of a global social plague, the truth is a slightly outsized ego has its benefits. In fact, numerous studies have found that people who see themselves as better than average are happier, more sociable, and often more physically healthy than their humbler peers. The swagger in their step is associated with a host of positive qualities, including creativity, leadership, and high self-esteem, which can propel success at work. Their rosy self-image imbues them with confidence and helps them endure hardship, even after devastating failure or horrific loss.

Bosnian War survivors provide a dramatic example. Psychologists and social workers who evaluated a group of survivors for depression, interpersonal difficulties, and other "psychological problems" found that those who considered themselves better than average were in better shape than those who had a more realistic view of themselves. A similar pattern emerged among survivors of 9/11. Feeling special seems to help survivors of tragedy face the future with less fear and greater hope.

The converse appears to be true as well: people who don't feel special often suffer higher rates of depression and anxiety; they're also less likely to admire their partners. It's not that their view of the world is wrong; very often it's more accurate compared to people who think highly of themselves. But they sacrifice their happiness for that realism; they see themselves, their partners, and the world itself, in slightly dimmer light. Researchers call this the "sadder but wiser effect."

It's ironic in a way, the reverse of what we've been taught about narcissism. It's not bad, but good to feel a little better than our fellow human beings, to feel special. In fact, we may need to. Where the trouble lies—whether narcissism hurts or helps, is healthy or unhealthy—depends entirely on the *degree* to which we feel special.

Narcissism, it turns out, exists on a spectrum. In moderation, it can, by inspiring our imagination and sparking a passion for life, open up our experience and expand our sense of our own potential. It can even deepen our love for family, friends, and partners. By far, the most powerful predictor of success in romantic relationships is our tendency to view our partners as better than they actually are. I call this "feeling special by association."

Psychologists Benjamin Le of Haverford College and Natalie Dove of Eastern Michigan University recently reviewed more than 100 studies involving nearly 40,000 people in romantic relationships and found that whether a couple stayed together beyond a few weeks or months depended most strongly not on partners having winning personalities, robust self-esteem, or feelings of closeness, but on one or both people holding *positive illusions*—that is, they viewed their partners as smarter, more talented, and more beautiful than they were by objective standards. Believing that we're holding hands with the most amazing person in the room makes us feel special, too.

But while moderate narcissism can enhance love, too much can diminish or even destroy it. When people grow dependent on feeling special, they become grandiose and arrogant. They stop thinking that their partners are the best or most important people in the room because they need to claim that distinction for themselves. And they lose the capacity to see the world from any point of view other than their own. These are the true *narcissists*, and at their worst, they also display two other traits of a so-called "dark triad": a complete lack of remorse and a penchant for manipulation.

Surprisingly, too little narcissism can be harmful as well. Remember Echo? She's the part of the myth we usually forget. She has no voice of her own. She's self-abnegating, nearly invisible. The less people feel special, the more self-effacing they become until, at last, they have so little sense of self they feel worthless and impotent. I call these people *echoists*.

Danger, then, lurks toward the ends of the narcissism spectrum. Only in the middle, where the need to stand out from 7 billion other humans doesn't blind us to the needs and feelings of others, lies health and happiness.

Another notion that we've mistakenly become wedded to is that our degree of narcissism is fixed throughout our lifetime. The fact is even healthy narcissism typically waxes and wanes, subsides and erupts, depending on our life circumstances and our age. When we're sick, for example, we normally move up the spectrum; we'll feel more deserving of others' time and care, even more entitled to it, than our healthier peers and family members. Similarly at work, when we feel the need to be recognized, admired, and appreciated—say when we're gunning for promotion—our narcissism spikes. In such instances, our hopes for the future ride on standing out from the herd. There are also specific life stages during which we need to see ourselves as special, such as pregnancy and adolescence; and others that move us toward Echo's end of the continuum, such as caring for a newborn or deferring our dreams to help support a partner's career. Both of these circumstances demand that we scale back our need to be in the spotlight.

But these peaks and valleys generally don't last forever. The crisis or transition passes and the drive to feel special returns to a healthy level. If we've moved closer to Echo's end of the spectrum, we find our voice again. And even if we've won the work promotion and quietly think we're better than our colleagues, the need to prove that to ourselves—and the world—isn't nearly as pressing. If it is, we're no longer in healthy territory.

Another common—and wrong—assumption is that damaging narcissists are always easy to spot. Yes, the loud, vain, self-aggrandizing ones who daily pop on our TV screens and stream through social media certainly are. They stick out like sore thumbs—which is probably a good thing; the truth is you'll find more narcissists in your life than echoists, and they'll be more of a concern (narcissists inflict damage on others, while echoists primarily hurt themselves). But not all narcissists advertise themselves so brazenly—some aren't even especially flashy or outgoing. And that makes recognizing them a lot harder.

There are also lower-profile *subtle* narcissists who are more difficult to detect, more common, and more likely to wreak havoc in our lives. They're the people we see every day: they're our lovers, spouses, friends, and bosses. Their unhealthy narcissism is often masked by their manner; they're often quiet, charming, capable of warmth, and even occasional empathy. Their signs are harder to spot—but they're still there. And if you're familiar with them, you can tease out the signals, including a tendency to flee emotions. In Chapter 7 we'll take a closer look at the signs that may be red flags, to help you evaluate your relationship with a subtle narcissist.

The idea that the person you sleep with or work beside might be a narcissist is shocking and depressing. Even more depressing is recalling the accepted wisdom that narcissism is a fixed personality trait or character flaw that never improves. But here, too, thinking has begun to shift. Many extreme narcissists do seem to be stuck (thankfully they're rare, only an estimated 1 percent to 3 percent of the US population). But some, milder narcissists may be able to change. Stripped down to its basic action, narcissism is a learned response, that is, a *habit* and, like any habit, it gets stronger or weaker depending on circumstances.

Narcissists bury normal emotions like fear, sadness, loneliness, and shame because they're afraid they'll be rejected for having them; the greater their fear, the more they shield themselves with the belief that they're special. Unhealthy narcissism isn't an easy habit to break, but people can become healthier by learning to accept and share the emotions they usually hide. And their loved ones can help them shift to the healthy center of the spectrum by opening up in the exact same way.

Just like most things in life, healthy narcissism boils down to striking the right balance. At the heart of narcissism lies an ancient conundrum: how much should we love ourselves and how much should we love others? The Judaic sage and scholar Hillel the Elder summarized the dilemma this way: "If I am not for myself, who am I? And if I am only for myself, then what am I?" To remain healthy and happy, we all need a certain amount of investment in ourselves. We need a voice, a presence of our own, to make an impact on the world and people around us or else, like Echo, we eventually become nothing at all.

We all sail between the Scylla of enervating self-denial and Charybdis of soul-killing self-importance. That's what narcissism is really all about—and you'll learn how to safely navigate the passage as we go along. But first, we have to untangle a mystery. If feeling special can be good for us, how on earth did we end up so obsessed with the idea that it's bad? Why are we so focused on the *dangers* of narcissism?

2

CONFUSION AND CONTROVERSY

HOW NARCISSISM BECAME A DIRTY WORD
AND WE FOUND AN EPIDEMIC

Many years ago, a close friend of mine, Tara, called me about an incident with her father and her two-year-old daughter, Nina. They'd been out for a stroll in the park when Nina suddenly became unglued, screaming and wailing to go home. Tara did what she could, but Nina remained inconsolable. After about half an hour, Tara announced, "We have to go, I'm sorry." Her father shot her a stern look, warning: "If you leave every time she pitches a fit, she'll think the world revolves around her!" Tara, fuming, fired back. "Yes. Yes, she will. And I think that's a *good* thing! Don't you?"

On the surface, this father-daughter quarrel was a generational battle over how to raise a child. But at a deeper level, their argument reflects two radically different views of human nature. Tara's dad seems to believe people are easily corruptible, requiring constant reining in to avoid becoming hopelessly self-centered, while Tara thinks we're all made of sturdier stuff and actually benefit from a little self-absorption now and then. The first position inevitably adopts a rather dim view of humanity, the latter a more optimistic one.

Without realizing it, Tara and her father had squared off in one of the oldest debates in history, one that's central to the confusion surrounding narcissism today.

THE BIRTH OF NARCISSISM

Long before the word *narcissism* had been coined, philosophers fought just as fiercely as Tara and her father over the place of the self in our moral priorities.

In 350 BC, Aristotle posed a question—"Who should the good man love more? Himself, or others?"—and answered it: "The good man is particularly selfish." In India two centuries earlier, the Buddha had spread the opposite view: The self is an illusion, a trick our minds play on us to make us think we matter. Buddhism suggested that this illusory self should never be our primary focus. Four centuries after Aristotle, Christian teachings added a negative fillip: making too much of oneself constitutes the *sin* of pride (and a quick path to hell). Excesses of the self underlie other sins—sloth, greed, gluttony, and envy—as well.

Down through the centuries, the debate raged, engaging philosophers from Thomas Hobbes (self-love is part of brutish human nature) to Adam Smith (self-interest benefits society, aka "greed is good"). It wasn't until the end of the 19th century, however, that the debate entered into the circles of medicine and psychology and the word *narcissism* first appeared. In 1898 pioneering British sexologist Havelock Ellis described patients who'd literally fallen in love with themselves, sprinkling their bodies with kisses from their own lips and masturbating to excess, as suffering from a "Narcissus-like" ailment. One year later, a German doctor, Paul Näcke, writing about similar "sexual perversions," coined the catchier term *narcissism*. But it was the founding father of psychoanalysis, Sigmund Freud, who in 1914 made the word famous in a groundbreaking paper: *On Narcissism: An Introduction*. He liberated the term from its sexual connotations (unusual for him), describing narcissism, instead, as a necessary developmental stage of childhood.

As infants, Freud wrote, we're convinced the world originates in us, at least all the exciting parts of it. We literally fall in love with ourselves, giddy with all the fascinating and sexy things we seem capable of. He called this stage "primary narcissism," and felt it wasn't just healthy, but also crucial to our capacity to form meaningful, close relationships. Our passion for ourselves as infants gives us the energy to reach out to others. We have to *overestimate* our own importance in the universe before we can see anyone else as important.

But Freud didn't know quite what to make of narcissism beyond infancy. Was it good or bad for adults? On the one hand, he felt that narcissism and love were closely linked; lovers often raise each other on a pedestal above the rest of humanity. He also pointed to charismatic leaders and innovators as proof that individuals who feel special can bring tremendous good to the world. But he was quick to condemn adult narcissism as well. If we don't let go of the childhood fascination with ourselves, he cautioned, it can lead to vanity (in his view found chiefly in women) and to serious mental illness, severing us from reality and turning us into delusional megalomaniacs. Freud's dual views on adult narcissism generated enormous confusion and set the stage for a crackling duel nearly fifty years later between two giants in mental health: Heinz Kohut and Otto Kernberg. Both men were born in Vienna to Jewish families and both trained as psychoanalysts. But they came of age under vastly different circumstances. Kohut, born in 1913, knew a Vienna full of hope and prosperity, brimming with rich artistic tradition and teeming with intellectual fervor. The advent of Hitler and the Third Reich changed all that. Soon after the annexation of Austria in 1938, Kohut fled his beloved city for England and then America, where he settled in 1940. Born in 1928, fifteen years after Kohut, Kernberg grew up in a grim and ominous Vienna in the shadow of encroaching Nazism. When he was 10 years old, he and his family fled to Chile, where Kernberg spent the next twenty years, far away from the home he'd once known; he moved to the United States in 1959. The two men's contrasting experiences seem to have colored their views of human nature. Darkness pervades Kernberg's view, while hope suffuses Kohut's.

THE RISE OF HEALTHY NARCISSISM

As a young psychoanalyst, Heinz Kohut, like Freud, quickly earned a reputation for brilliance as a clinician, researcher, and teacher/lecturer. (He was renowned for his ability to commit entire therapy session transcripts to memory and to deliver compelling talks without a single note to prompt him.) Throughout most of his career, he remained one of Freud's staunchest defenders. But in the 1970s, he split from the orthodox Freudian community to found an entirely new school of thought, Self Psychology, devoted to understanding how people develop a healthy (or unhealthy) self-image.

Kohut believed that Freud had stumbled by placing sex and aggression at the center of human experience. It's not our baser instincts that drive us, Kohut argued; rather, it's our need to develop a solid sense of self. And for that, he said, we don't just need other people; we *need* narcissism. Freud had all but elevated self-reliance to the level of virtue. We should be fully autonomous as adults, declared the master, demanding neither approval nor admiration. But where Freud saw narcissism as a mark of immaturity, an infantile dependency to be outgrown, Kohut saw it as vital to well-being throughout life. Even as adults, we need to depend on others from time to time—to look up to them, to enjoy their admiration, to turn to them for comfort and satisfaction.

Young children only feel like they matter—only feel like they *exist*—when their parents make them feel special. Parents who pay attention to their children's inner lives—their hopes and dreams, their sadness and fears, and most of all their need for admiration—provide the "mirroring" necessary for the child to develop a healthy sense of self. But young children also need to idolize their parents. Seeing their mother and father as perfect helps them weather the storms every fledgling self goes through as we face life's inevitable disappointments. *I'm awesome anyway*, you can tell yourself when bullied at school or flunking math, *because my parents think so. And my parents are perfect, so they should know.*

Kohut believed that children gradually learn that nothing—and no one—can be perfect and so their need for self-perfection eventually gives way to a more level-headed self-image. As they witness the ways healthy adults handle their own flaws and limitations, they begin coping more pragmatically, without the constant need for fantasies of greatness or perfection. At the end of their journey, they acquire healthy narcissism: genuine pride, self-worth, the capacity to dream, empathize, admire and be admired. This, Kohut said, is how any of us develops a sturdy sense of self.

But when children face abuse, neglect, and other traumas that leave them feeling small, insignificant, and unimportant, they spend *all* their time looking for admiration or finding people to look up to. In short, Kohut concluded, they become *narcissists*—vulnerable, fragile, and empty on the inside; arrogant, pompous, and hostile on the outside, to compensate for just how worthless they feel. People, in their eyes, become jesters or servants in their court, useful only for the ability to confirm the narcissist's importance.

The rest of us, if our parents do their job right, never lose our moments of grandiosity. Nor should we. In Kohut's eyes, it was madness to think of lofty dreams as inherently bad. If anything, they provide a depth and vitality to our experience, fueling our ambitions and inspiring creativity. Composers and artists throughout history, he noted, often have moments of self-importance. To produce anything great—to even sit down and try—often requires feeling that we're capable of greatness, hardly the humblest state of mind. Kohut refused to see some of civilization's greatest creations simply as the result of illness. Instead of stamping out narcissism, he argued, we should learn to enjoy it as adults. Narcissism only becomes dangerous, taking us over and tipping into megalomania, when we cling to feeling special like a talisman instead of playing with it from time to time. It all depends on how completely we allow grandiosity and perfectionism to take us over.

There's an appealing romanticism to Kohut's vision of narcissism. It allows us to disappear into ourselves, like Narcissus diving into the pool, but instead of drowning and becoming lost forever, we discover another world, richly populated with shimmering versions of everyone we love. Once there, we, too, take on a kind of otherworldly glow. For a time, we're different, special, set apart from the rest of humanity. If we're healthy enough, we can reemerge and rejoin the ordinary world, bringing our bounty, such as empathy and inspiration, with us. Where Freud's narcissist is childish—a Peter Pan figure stubbornly refusing to become an adult—Kohut's is, at his best, an adventurer, slipping in and out of intoxicating dreams of greatness.

By the 1970s Kohut's self-psychology movement had become something of a juggernaut and his views on narcissism had become widely accepted. In fact, when the third edition of the *Diagnostic and Statistical Manual* (*DSM*)—the guide to classifying mental disorders published by the American Psychiatric Association—hit the shelves in 1980, it carried a brand-new description of unhealthy narcissism very similar to the one Kohut had proposed. By then many mental health experts believed feeling special could lead to many good things—and the dangers, while very real, had been overstated. But the tide was about to change.

THE RISE OF THE DARK NARCISSIST

Otto Kernberg agreed with Kohut that healthy narcissism provides us with self-esteem, pride, ambition, creativity, and resilience. But he diverged sharply with Kohut's theory when it came to unhealthy narcissism. Whereas Kohut viewed even grandiose narcissism in a somewhat benevolent light, Kernberg saw it as inherently dangerous and harmful.

Likely due to his exposure at an impressionable age to Nazism and Hitler (one of the most dangerous megalomaniacs who ever lived), Kernberg believed in the presence of evil in the world. His experience during psychoanalytic training reinforced his dark views of human nature—Kernberg cut his teeth professionally working in hospitals and clinics with severely mentally ill patients prone to aggression and psychosis, while Kohut arrived at his theories treating privileged patients in his luxurious private offices. In Kernberg's view, narcissists, at their most destructive, are masses of seething resentment—Frankenstein's monsters, crudely patched together from misshapen pieces of personality. They'd been failed so horrifically as children, through neglect or abuse, that their primary goal is to avoid *ever* feeling dependent again. By adopting the delusion that they're perfect, self-contained human beings (and that others are beneath them), they never have to fear feeling unsafe and unimportant again.

Far more loyal to Freud's legacy than Kohut, Kernberg refused to abandon the idea that sex and aggression fueled much of our behavior. Like Freud, he saw human beings as roiling cauldrons of hostility and lust, driven by their darkest and often cruelest passions. The most dangerous narcissists, in Kernberg's view, may even be born with too much aggression wired into them; they're frightening mutations, given to a far stronger impulse to envy, attack, and destroy their fellow human beings when they feel hurt. Made to feel worthless as children and fueled by their overabundance of hate, they ravage the rest of humanity out of revenge, using people to satisfy their own needs and casting them aside when they're done. Kernberg called the most frightening of these specimens "malignant narcissists."

The only sensible response to this threat, according to Kernberg, is to dismantle the warped self-image and reconstruct it in more benevolent form. He believed that narcissists were capable of reform and that confronting them with the truth of the danger they pose is the first step in changing their behavior. We certainly can't stop the threat of destructive narcissism by *feeding* their need to feel special. That's a bit like letting the monster loose to terrorize the villagers. This was anathema to Kohut, who advocated approaching narcissists with empathy. They need our understanding, he said, if they have any hope of getting better. Kernberg, still allied with Freud's bleak vision of humanity, could only see Kohut's stance as dangerously naïve.

Kohut's and Kernberg's competing theories were battled over through conferences and papers, with neither side gaining ascendancy. But after Kohut succumbed to cancer in 1981, Kernberg was left alone in the spotlight and his views, particularly of malignant narcissism, spread widely. They were helped into public consciousness by historian and social critic Christopher Lasch's popular 1979 book, *The Culture of Narcissism*, which drew heavily on Kernberg's frightening image of destructive narcissism. In most people's minds, narcissism became synonymous with *malignant* narcissism.

This image began to take hold, magnified by the idea that narcissists weren't rare creatures that we had only the slightest chance of encountering in our lifetimes, but monsters standing on every street corner, sitting in the next cubicle, and sleeping in our beds. And soon one little test enabled the paranoia to spread like wildfire.

AN EPIDEMIC OF NARCISSISM—
OR A LITTLE MEASUREMENT MAGIC

Introduced in 1979, the Narcissistic Personality Inventory (NPI) is a basic tool of psychology researchers, and is routinely administered to undergraduate psychology students in the United States and around the world. (If you ever studied psychology in college, you probably took the NPI.) Respondents read 40 paired statements and check off which one of the two best describes themselves. For example: "I like to show off my body" and "I don't particularly like to show off my body" or "I find it easy to manipulate people" and "I don't like it when I find myself manipulating people." Each narcissistic choice gets one point; the opposite choice gets a zero. Points are added up and people who score well above average earn the title of narcissist.

In 2009, thirty years after the inventory's start-up, psychologist Jean Twenge, of the University of Texas, compared average totals by year for thousands of US students and announced that the averages had risen "just as fast as obesity from the 1980s to the present." She proclaimed that a "narcissism epidemic" is raging among millennials—and underscored her contention by using the same shock phrase for the title of her book. *The Narcissism Epidemic*, coauthored with psychologist Keith Campbell, of the University of Georgia, explored the alleged rampant arrogance and entitlement of today's youth. This was the dramatic follow-up to her first book, *Generation Me*, in which she declared, based on the same research, that "today's young Americans are more confident, assertive, entitled—and more miserable than ever before."

Twenge placed the blame for this epidemic squarely on shoulders of parents and educators who made a generation of children coming of age in the 1980s and '90s feel, perhaps, a little *too* special. After all, it had become commonplace for classrooms to be plastered with positive-reinforcement posters proclaiming things like "You are unique!"; for trophies to be handed out for effort, not accomplishment; for parents to remind their children at every turn that they were perfect just as they were. Love yourself enough, the message seemed to be, and you can do anything. Some educators even argued that boosting self-esteem would be something of a panacea, promoting well-being and happiness, preventing bullying—possibly even reducing crime. Make kids feel special, they argued, and great things will follow.

While this self-esteem campaign doesn't appear to have had a positive impact on crime rates, bullying, or achievement scores,

Twenge argued that it did have a significant cultural impact: it created "an army of narcissists." In an effort to help children feel better about themselves, we'd inadvertently ruined them. Having given them too much leeway and swollen heads, we hadn't simply damaged our kids; we created a generation that posed a threat to the entire world.

Twenge's theories touched a cultural nerve. The press was already rife with reports of overinvolved parents who coddled their children, chewing out their sons' or daughters' teachers for dishing out bad grades or calling during job interviews to speak to their prospective employers. Headlines buzzed with shocking tales of millennials' sense of entitlement: disgruntled administrative assistants who slacked off at work, convinced that secretarial duties were beneath them; entry-level workers who held court when they should have been listening to their boss; new hires who spent entire meetings glued to their smartphones, texting friends instead of taking notes. And now, it seemed, Twenge had provided an explanation for all the bad behavior.

Her conclusions, however, have drawn fire right from the start—and the evidence she marshals to support the idea of a narcissism epidemic has come under the heaviest attack. The NPI, on which Twenge draws so heavily, is a deeply flawed measure. Under its design, agreeing with statements that reflect even admirable traits can inch people higher up the narcissism scale. For example, picking "I am assertive" and "I would prefer to be a leader" counts as unhealthy even though these qualities have been linked repeatedly in decades of research to high self-esteem and happy relationships. People who simply enjoy speaking their mind or being in charge are clearly different from narcissists who enjoy manipulation and lies. But the NPI makes no distinction. More people checking these salutary statements could easily account for millennials' rising NPI scores through the years, and that's what some studies indicate has happened.

Second, numerous large-scale studies, including one of nearly half a million high school students conducted between 1976 and 2006, have found little or no psychological difference between millennials and previous generations (apart from a rise in self-confidence). In fact, one study of thousands of students suggests that millennials express greater altruism and concern about the world as a whole than do previous generations, prompting psychologist Jeffrey Arnett, of Clark University to call them "GenerationWe." The results of a 2010 Pew Research Report, surveying a nationally representative sample of several thousand millennials, also stands in stark contrast to Twenge's findings. Millennials, the Pew authors concluded, get along well with their parents, respect their elders, value marriage and family far over career and success, and are "confident, self-expressive, and open to change"—hardly the portrait of entitled brats.

But there's another far more troubling problem with using the NPI to declare an epidemic: we have no way of knowing whether or not people scored as "narcissists" remain so over time. No study has followed up on these thousands of college students after they graduated. Furthermore, just about every theory of adolescence and early adulthood presumes that the young are only temporarily a self-absorbed bunch, and research seems to support that view. We used to think that was a good thing: the bright-faced idealism of youth. The young believe themselves capable of anything; they're ready to take over the world and make it a better place. Most of us, in our less cynical moments, appreciate their ostentatious energy. But just as with other temporary bouts of narcissism brought on by specific life stages, such enthusiasm eventually fades. As we approach our thirties, most of us come back down to earth, and our self-importance, and yes—self-absorption—give way to the realities of life.

Though we currently seem obsessed with Kernberg's dark narcissism, the pervasive better than average effect—where healthy people do appear to feel special—suggests that Kohut's benign view is the right one.

We need our grandiosity at times to feel happy and healthy. And a growing body of recent research concludes that a little narcissism, in adolescence, helps the young survive the Sturm und Drang of youth; moderate teenage narcissists are less anxious and depressed and have far better relationships than their low and high narcissism peers. Likewise, corporate leaders with moderate narcissism are rated by their employees as far more effective than those with too little or too much. And my own research with my colleagues is pointing in the same direction: only people who never feel special or feel special all the time pose a threat to themselves and the world.

The difference between narcissists and the rest of us is one of degree, not kind. To better understand that, we need to explore the full range of the narcissism spectrum.

3

—

FROM 0 TO 10

UNDERSTANDING THE SPECTRUM

When my daughters were in kindergarten, they loved to visit the Cambridge Museum of Science. One exhibit, in particular, fascinated them. It consisted of a small tile with a lamp shining down on it. By turning a knob on the lamp, they could change the color of the light. But each time the lamp changed color, so did the tile. What seemed to be a bright red tile, a few moments ago, would deepen into purple, then turn yellow, then green, and on and on. At the edges, some colors would blend, making it hard to discern any one color at a time. A seemingly trivial question, *What color is the tile?*, suddenly became far more complicated.

We tend to like clear, distinct categories—it makes life easier to impose order on the world. The tile is either green or red, but it can't be both. Similarly, we like to think in stark extremes— full or empty, black or white, good or bad. But as soon as we start looking more closely at our world, the categories blur. Even the paint on our walls seems to change color throughout the day, depending on the directness and intensity of the light. There are gradations and nuance to almost everything in life, including attitude, emotion, and personality.

So instead of regarding narcissism in all or nothing terms, imagine a line stretching from 0 to 10, like the one below, with the desire to feel special slowly growing as we move from left to right.

THE NARCISSISM SPECTRUM

0	1	2	3	4	5	6	7	8	9	10
abstinence		*habit*		*moderation*			*habit*		*addiction*	

Life at either of the extremes, whether at 0 or 10, isn't a particularly healthy place to be. At 0 people *never* enjoy feeling special in anyway. Perhaps they never have. At first, this might sound healthy. Most of us have it drummed into our heads, whether by religion or family or culture, that anything even approaching the desire for special treatment or attention is *bad*. Our distaste is epitomized by the question *What makes you so special?* We all recognize the reprimand in the rhetoric. What people really mean is *You're acting like you're special. Stop it!* In most cultures around the world, selflessness is often held up as the ultimate virtue. No one has a right to feel special anyway, the argument goes, so we should celebrate people who never indulge the feeling.

But bear in mind what that really means: unrelenting selflessness, feeling abjectly ordinary, no more deserving of praise or love or care than anyone regardless of the circumstances. It doesn't take long to see that this presents a range of problems. Say, for example, you've lost your beloved mother to a horrific car accident. Most people would agree that you deserve special attention; during grief, our pain should take center stage for a time. Living at 0 means you not only wouldn't accept sympathy and assistance, you might even actively push it away. I once worked with a woman who rigidly refused to let anyone help or support her, even after her husband died. "Please—don't trouble yourself," she'd say when anyone tried to pick up groceries for her or drive out to visit her (she lived an hour from most of her friends). She was determined to be alone instead of surrounded by supportive companions giving her special attention.

Life at the far right is just as bleak. While people at 0 assiduously avoid the spotlight, those at the far right either scramble for it or silently long for it. In their minds, they cease to exist if people aren't acknowledging their importance. They're *addicted* to attention, and like most addicts, they'd do anything to get their high, so even authentic love takes a backseat. At 10 our humanity collapses under the weight of empty posturing and arrogance. Think of Bernie Madoff, who swindled hundreds of millions of dollars from his clients and who, when caught, scoffed at the "incompetence" of the investigators for not asking the right questions. Even as he faced life in prison, he still managed to feel superior.

Being at 1 or 9 isn't much better. People at 9 are still in the territory of dark narcissism; they can live without elbowing their way into the spotlight, but it pains them to do so—so much so that they need professional help to break the habit. (Think of Don Draper of the TV series *Mad Men*, hopping from affair to affair, desperately seeking excitement and attention; he can't stop even after he sees the damage his lies and infidelity have inflicted on his family.) People at 1 suffer just as much; their aversion to feeling special is unyielding. They might tolerate a little attention on birthdays, but they hate it.

As we approach 2 and 3 and 7 and 8 on the spectrum, we

leave behind the compulsive rigidity found near 0 and 10, and enter the area of *habit*. There's greater flexibility of feeling in this range, and therefore, more possibility for change. On the left, at 2, people enjoy feeling special, albeit infrequently; at 3 they may secretly dream of greatness. On the right, at 8, they might occasionally set aside their flamboyant dreams and devote some thought to other people; at 7, they've begun to show signs of humanity again, including the occasional ability to admit to ordinary faults.

A hallmate of mine in college offers a good example of someone around 3 on the spectrum. She enjoyed birthdays and accepted compliments, but she still hated it when anyone tried to take care of her. She'd actually get up and clean dishes as soon as someone tried to clear hers. She struggled with this inability to let others do things for her, confessing to me late one night, "I hate that it's so hard for me to accept help or special treatment." Likewise, a dormmate of mine who lived at 7 felt self-conscious about the way he'd name-drop or find a way to work his high test grades into casual conversations. "I know it's wrong," he said, "but I do it so people will be impressed. I'm worried that if I don't, they won't think much of me at all." Habitual echoists and narcissists recognize that their behavior might be less than healthy; they just can't always keep it in check.

The healthiest range is found in the center, at 4 through 6; it's the world of moderation. Here, we might find intense ambition and occasional arrogance, but feeling special isn't compulsive anymore. It's just fun. At 5, in the very center, there's no relentless need to feel—or avoid feeling—special. People here enjoy vivid dreams of success and greatness, but don't spend all their time immersed in them. You'll notice that 6, though it tips past the center, is still in the healthy range. That's because it's quite possible to have a strong drive to feel special and still remain healthy. Healthy narcissism is all about moving seamlessly between self-absorption and caring attentiveness—visiting Narcissus's shimmering pool, but never diving to the bottom in pursuit of our own reflection.

WIGGLE ROOM:
MOVING UP AND DOWN THE SPECTRUM

Recently, I got slammed with a miserable cold, one that left me feeling grumbly and demanding. I just wanted someone to take care of me. But then a friend called who'd just lost his job, forcing him to uproot himself and find work in another part of the country. Suddenly, my cold wasn't so important anymore. I rose from bed, cleaned myself up, and went to talk with him.

Most models of human behavior consider flexibility to be the hallmark of mental health. We adapt our feelings and behavior to fit the circumstance. When it comes to narcissism, similarly, only the most extreme echoist or narcissist becomes fixed at one end of the scale. Healthy people generally remain within a certain range on the spectrum, moving up or down a few points throughout their lives. Nevertheless, we're all prone to climbing even higher on the scale if something provides a big enough push.

Narcissism spikes dramatically, for example, when we feel shaky about ourselves: lonely, sad, confused, vulnerable. In adults, major life events like getting divorced or becoming sick in old age often trigger a large surge of self-centeredness as we struggle to hold on to our self-worth. In younger people, narcissism tends to peak during the teen years. Adolescents often betray a staggering sense of omnipotence, as if they're somehow above natural and man-made laws (fatal accidents might happen to others who drive drunk, for instance, but certainly never to them). Teens are well known for elevating even the act of suffering to great heights—prone to fits of despair, convinced no one can fathom the pain of their unrequited crush, or the searing humiliation of not owning the next cool smartphone. Nothing else—and often no one else—matters more than the anguish they feel.

Though vexing for parents, this adolescent peak in narcissism is normal and understandable. This is the time when we develop an individual identity, separating from our parents to become our own person. We push away from people who've held sway over us, even though we know, somewhere deep inside, that we aren't yet equipped to handle the world on our own. It's at times like these—when we need people but aren't sure if we can or should have their support—that we lean heavily on feeling special. It boosts our confidence, however temporarily. And while it's not genuine or lasting self-assurance, it gets us through a rough time. Once we're through adolescence, narcissism falls sharply; it's time to get on with the business of adulthood—and that means thinking about people other than ourselves.

VARIETIES OF SPECIAL: EXTROVERTED, INTROVERTED, AND COMMUNAL NARCISSISTS

You've no doubt come across extroverted narcissists. That's the kind of narcissist you're used to hearing about, the one about whom all the fuss is made. They're loud, vain, and easy to spot. They flaunt their money and possessions, scramble to be the center of attention at every occasion, ruthlessly jockey to rise through the ranks at their office. But narcissism manifests itself in other ways, as well. An intense drive to feel special can yield two other types of narcissistic behavior: introverted and communal.

Introverted narcissists (also called "vulnerable," "covert," or "hypersensitive" in scientific literature) are just as convinced that they're better than others as any other narcissist, but they fear criticism so viscerally that they shy away from, and even seem panicked by, people and attention. Their outward timidity and wariness makes them easily mistaken for self-effacers at the far left of the spectrum. But what makes them different from echoists is that they don't feel inferior. They believe they harbor unrecognized intelligence and hidden gifts; they see themselves as more understanding of, and more attuned to, the intricacies of the world around them. In self-report, they agree with such statements as *I feel that I am temperamentally different from most people.* To an observer, these people appear fragile and hypersensitive. In conversation, they're apt to jump on a misplaced word, or a change in tone, or a brief glance away, and demand *What did you mean by that?* or *Why are you turning away?* There's an angry insistence to introverted narcissists: they seethe with bitterness over the world's "refusal" to recognize their special gifts.

Communal narcissists, a type more recently identified by researchers, aren't focused on standing out, being the best writer or most accomplished dancer or the most misunderstood or overlooked genius. Instead, they regard themselves as especially nurturing, understanding, and empathic. They proudly announce how much they give to charity or how little they spend on themselves. They trap you in the corner at a party and whisper excitedly about how thoughtful they've been to their grieving next-door neighbor: *That's me—I'm a born listener!* They believe themselves better than the rest of humanity, but cherish their status as *givers*, not takers. They happily agree with such statements as *I am the most helpful person I know* and *I will be well known for the good deeds I have done.*

As you can see, not all narcissists look and sound alike and, no doubt, we'll discover even more than these three variations over time. But remember—regardless of their differences, they all share one overriding motivation: each and every one of them desperately *clings* to feeling special. They just do it in different ways.

SPECIAL DEMOGRAPHICS:
AGE, GENDER, CAREER

As you've learned already, narcissism may come more easily to the young; people under 25 tend to be the most narcissistic, with the drive to feel special declining as we age. But what about that perennial question of who's more narcissistic—men or women? Most studies only capture the extroverted narcissists and, when it comes to this group, researchers consistently find slightly more men than women in the mildly unhealthy range (7 to 8, by this book's scale). In stark contrast, as soon as we get to the extreme right of the spectrum, men dominate sharply; they're double the number of women.

This difference is at least partly attributable to gender roles. In most societies, women are criticized for being loud and assertive, while these same qualities are encouraged in men. So it's no surprise there's a slight difference in habitual narcissism and a huge difference in the addictive kind. It's one thing for a woman to be extremely confident and hypercompetitive, but being floridly arrogant and forceful departs dramatically from common notions of how women should behave.

Research on communal narcissism is just beginning to get under way, but so far, it seems to affect men and women in

equal numbers. Communal narcissists can either quietly believe they're the best parents or friends or humanitarians in the
world or get up on stage and announce it to everyone. With men
outnumbering women in the loud camp and women edging
past men in the quieter one, the gender difference washes out.
Interestingly, introverted narcissists, too, are about equally di-
vided between the sexes.

Some professions seem to be magnets for people from certain regions of the spectrum. People on the high end of the
spectrum tend to gravitate toward careers where there's an
opportunity for power, praise, and fame. US presidents are
more narcissistic, on average, than most ordinary citizens, according to psychologist Ronald J. Deluga, of Bryant College,
who used biographical information on every commander in
chief from George Washington through Ronald Reagan to
score them on the NPI. Predictably, high-ego presidents like
Richard Nixon and Ronald Reagan ranked higher than more
soft-spoken leaders like Jimmy Carter and Gerald Ford, but
almost all presidents scored high enough to be considered
"narcissists."

Psychologists Robert Hill and Gregory Yousey, of Appa-
lachian State University, also studied the narcissistic tendencies of politicians (excluding presidents), comparing them with
librarians, university professors, and clergy. Politicians again
ranked higher in narcissism than any other group. Clergy and
professors were deemed the healthiest, with librarians the least
narcissistic. Unlike the politicians, none of the other profes-
sionals scored high enough to earn the label *narcissist*, though
librarians certainly scored low enough to flirt with echoism.

The performing arts is an arena with a heavy pull for narcissists—no surprise there; it's *show* business, after all—but
here, too, there are shades of narcissism if you look closely
enough. Dr. Drew Pinsky, host of the radio show *Loveline*, did

just that, by asking every celebrity who appeared on his show
to take the NPI. Then he and psychologist S. Mark Young, of
the Marshall School of Business at the University of Southern California, compared the actors' scores to those of people
in other artistic areas. Actors and comedians, it turned out,
fell near the middle of performers in narcissism (women were
more narcissistic than men, possibly because their appearance
is more important to their success). Musicians were the least
narcissistic. And the most narcissistic? (Drumroll . . .) Reality TV stars. Based on the data, Pinsky and Young concluded
that all the celebrities started out high in narcissism, which,
in turn, probably drew them to their flashy careers. For the
record, Pinsky and Young also looked at MBA students for
comparison, since they often score higher than other groups
in narcissism—but the celebrities still won.

Few of us regularly interact with heads of state, celebrities,
or even MBA students, so the narcissism we're most likely to
encounter will be in the people we see regularly—our relatives,
friends, colleagues, dates, and mates. What does that look like?
Let's start with ordinary folks at the extreme ends of both sides
of the spectrum.

LIFE AT 2: SELF-DENYING

Sandy, 28, is single and works as an administrative assistant at
a biotech firm. She came to see me after a recent upset at work.
Her boss had decided to throw a party in her honor—his way
of saying thanks for her tireless effort to make the company's
past year especially prosperous.

"He was giving me an office MVP award and the day he selected for the party was also my birthday so he decided to kill
two birds with one stone." She grimaced as she spoke and her

thin frame seemed to shrink further in her loose black pantsuit. "My boss had spent a lot of time setting it up as a surprise, but I kind of figured out what was happening. People whisper around the coolers." Unhappy with the party, Sandy tried to get it canceled. "I told my boss's partner I'd been having trouble concentrating at work because I kept feeling awkward and anxious thinking about it. I managed to get it called off."

"What made you so uneasy?" I asked

"I can't stand compliments. They make my skin crawl. I've never liked being the focus of anything. I don't like birthday parties, either, let alone surprise ones."

"Any idea why?"

"No clue," she said. She stared at a large blue and green abstract painting on the wall in front of her. "All I know is I feel uneasy. I don't like people fawning over me."

Though Sandy was nearly allergic to gratitude from others, she had no trouble lending friends her support. But here, too, when they showed their appreciation with flowers or cards, she was visibly uncomfortable and accepted their tributes reluctantly.

"How about from your boyfriend?" She'd been living with Joe for three years in a small apartment just minutes from her office.

"I can't stand it when he compliments me or tries to take care of me." She squirmed, shifting back and forth in her seat. "I tell him he doesn't need to. I'm not a little kid."

Her evident distress had begun causing ripples in her relationships at work, at home, and with friends. "My boss was hurt. He said he just wanted to do something special for me." Joe, too, had clearly grown weary of such a one-sided relationship. "He got really angry the other day because he just wanted me to tell him which restaurant I preferred for my birthday dinner. I was tired of talking about it." She frowned. "I told him, 'Why don't we just stay home and cook—or you can pick wherever you want—it's up to you.' "

Joe had thrown up his hands in disgust, and growled, "You never let me *do* anything for you!"

"That's the problem," I said. "Sometimes people need us to be able to take center stage. It helps them feel special, too."

Sandy is a great example of the dangers of living at around 2 on the spectrum. The people who dwell there aren't just unfamiliar with feeling special, they're afraid of it.

Most of us feel a little boost when we receive praise and attention for our accomplishments. For a time, the spotlight has shifted to us. But for people near 0—extreme echoists—even positive attention can be terrifying. It's not necessarily because they feel ashamed or defective, though some might. It's just that they're convinced that being ordinary is the safest way to live. They stay in the shadows because, as the Japanese saying goes, "The nail that stands out gets pounded down." Even more, they dread becoming a burden. This isn't the feigned concern of martyrs who proclaim, "I don't want to put you to any trouble," while loudly voicing complaints that demand everyone's attention—this is real fear.

People like Sandy worry so deeply about seeming needy or selfish that it's often difficult for them to recognize they have any needs at all. It's also exhausting working so hard to expect nothing at all, which is why people at this end of the spectrum can lapse into confusing bouts of sadness. They feel depleted, but what they need to replenish themselves is buried so deep they're not even sure how to ask for it.

The most common feature of echoists is a deep dread of becoming narcissistic in any way. They're constantly on guard for any signs of selfishness or arrogance in themselves, so much so that they can't even enjoy being doted upon. Their vigilance comes with a steep price. People feel closer to us when we allow ourselves to become a gleam in their eye. Enjoying our moments on the pedestal elevates not just us, but also those we love.

LIFE AT 9: SELF-SERVING

Gary, 24, single, is a business school student who was referred to me by his dean, an old friend of his parents, who'd grown concerned and irate about his absences from class.

"I've got bigger fish to fry than going to class," Gary told me, smiling broadly. "I'm starting up a company with a friend. We got the idea one night when we'd been drinking for hours. But it's a great plan." He'd arrived ten minutes late to my office, but didn't seem the least bit contrite about being tardy. "Just came from an investor meeting," he'd explained, grasping my hand firmly in greeting.

"Terrific," I responded. "Congrats."

"I know how to sell myself," he said, shrugging. "It's what I do."

I could see what he meant. Sitting in a classic power position—arms clasped behind his neck, elbows out—he looked more like a business executive than a student. He dressed the part, too—a sleek navy blue suit, gleaming leather shoes, a red-and-blue striped tie.

"Are you any good at this?" he asked. "I don't have much time to waste."

"Guess we'll find out," I said, feeling sure he'd already decided. "As I understand it, you might get kicked out because you've missed so many papers and assignments."

"Dean tell you that?" he shot back, snidely. He leaned back, crossing his arms. "Listen, they have to keep me in school. I might be the best thing that's happened to them in a while. The least they can do is try to hold on to me. If they don't, they'll see what a mistake they've made when my idea takes off and I make a killing."

"You can appreciate the dean's position, though?" I asked, curious if he had any perspective on how much jeopardy he'd placed himself in.

"I can talk my parents into anything," he assured me. "I can talk pretty much anyone into anything," he added. "They'll convince him just like they did before." He combed his fingers though his hair. "People are making a big deal out of nothing. I can crank out the rest of my work, no problem."

"What made you decide to come to see me?" I asked. "You didn't have to."

"I figured you just need to give me a clean bill of health," he answered matter-of-factly.

"Ah," I said. "It doesn't quite work that way, unfortunately. We need"—here he cut me off.

"Look," he said, "I get that I have to convince the dean's bosses. That's why my parents are paying for this. If you can't help me, I'm sure I can find someone else to get the job done." He started getting up to leave.

"You can leave," I said. "But part of the problem is you don't think you need *anyone's* help. You've got a lot of talent and ambition, which is fantastic. But you can't rely on that alone to carry you. If that worked, you wouldn't be sitting across from me now. And the dean wouldn't be meeting with the school next Monday about whether or not this is your last semester there."

That seemed to get his attention. He sat back down.

This is the face of narcissism we all know and loathe: arrogant, entitled—at times frightening. People at 9, extreme narcissists, often think themselves above normal rules and expectations. Whatever they're paid, it's not enough. Whatever wrongs they commit against others, they're explained away. It never occurred to Gary for an instant that he might really be kicked out. Mysteriously, he believed that the university needed him far more than he needed it. He was convinced that his talent as an entrepreneur would save him.

People who live at 9 or 10 cling to their special status for dear life. Their belief that they're somehow above the rest of us mere mortals might even reach delusional levels, as it did for Gary, who honestly felt that he could do whatever he wanted and still remain in school. This sense of being a "special exception" also explains many other characteristics of people who live on the far right—becoming angry at the smallest slights, willing to do whatever it takes to get what they want, seeing other people as extensions of themselves.

Extreme narcissism blinds people to the feelings of others. That's one of the reasons we find it so unpleasant to be around people at this high end of the scale. The men and women who live near 10 are too preoccupied with their own need to be recognized and rewarded to consider the needs of other people.

Gary's parents had been on the phone with him nightly for a week, urging him to seek help. "I'm at my wit's end," his mother had said, in a tearful message left on my voice mail. Gary shrugged it off. "She gets that way." The dean had been a staunch defender, despite Gary's blithe attitude about his imminent expulsion. He'd known Gary since he was a toddler and clearly thought of him as a son. The whole situation had obviously been taking its toll—the dean sounded exhausted in his messages. But Gary seemed oblivious to just how anxious he'd made everyone around him, especially those who cared about him. "The dean's as big a worrier as my mom."

Those on the far right tend to regard others as tools for their personal use. Gary treated me, from the start, like a simple-minded servant. He quickly turned on me when I told him I couldn't just write a letter telling the administration he was fine.

Gary also had no insight into his problems. When feeling special becomes an addiction, there's no room to acknowledge flaws, no matter how obvious they are to everyone else. People like Gary are notoriously bad partners and friends. Their lack of empathy hobbles them relationally, leading to frequent lies and infidelity. But people who live around 9 don't see it. In fact, ask them if they're comfortable with deeper intimacy, capable of sharing sadness and loneliness with those they care about, and they'll often say they're good at that, too. They have such little self-awareness they can't even recognize the limits of their ability to love.

LIFE AT 5: SELF-ASSURED

Lisa, 41, married, Asian American, is the executive director of a nonprofit that serves the local Asian community. She came to me after her mother died from a massive stroke. "She didn't even make it to the hospital," Lisa told me in our initial phone call. "I've been different lately, a little off my game, so I thought I should speak to you."

When I met Lisa in the waiting room, she was chatting with another therapist's client (I'm in a suite of offices with other therapists). I'd seen this other woman before, but I'd never seen her speak to anyone. She usually sat quietly, reading a magazine or scrolling through her smartphone. Today she was smiling.

"Nice meeting you," said Lisa, as she waved goodbye to the woman. And I could tell she meant it.

I led Lisa down the hall. Before she sat, she smoothed out

her skirt—navy blue and business length, with a matching suit jacket—and adjusted her ponytail. "I'm a big believer in staying on top of things. I don't want this—whatever this is—to get out of hand."

Since her mother's death, Lisa had thrown herself into a bunch of new projects. She was so tightly scheduled she barely had time to think. "I'm always on the go," she said. "But I'm really pushing myself these days."

Lisa, who had successfully launched a number of programs for the homeless and elderly, was something of a local celebrity. She had myriad political connections, from alderman to senators, and made frequent TV appearances. "Most people hate all the media work, but I love making speeches or being on camera. I feel so alive then. I'm kind of a ham, anyway. I used to be an actress." She'd hit the stage as a toddler and continued acting in plays and musicals through high school. "I adore applause."

"But lately it feels like too much?" I asked.

"Isn't it?" she asked, and took a deep breath. "How do you know when it's healthy—all this chasing after success? All these big dreams?" I could tell she'd gotten to the heart of what had been eating at her. She visibly relaxed once she'd said it, her eyes glistening.

"You've been more driven than usual these days, since losing your mother. We can work on that. But the joy you take in dreaming big hasn't just made you happy—it's made others happy, too," I said. "I'd say that's the definition of health."

At the heart of healthy narcissism is the capacity to love and be loved on a grand scale. People who live in the center of the spectrum don't always take to the stage, but when they do, they often lift others up with them.

Lisa embodied many of the traits of healthy, centered narcissism. Her grief had driven her into the public eye a little more than usual, but she had enough self-awareness to realize something was wrong. People who live in the center know when their grandiosity is getting the better of them. They know when they're getting too caught up in themselves. Lisa's delight in feeling special never blinded her to how other people felt. Her main concern came down to her husband, Doug. She worried he'd become lonely—and he probably had.

"I found him in front of the TV the other day," Lisa admitted, "and he was looking pretty down. I'd been up all night working on a project and hadn't come home."

That prompted a long conversation in which Doug admitted to Lisa that he felt she'd been too self-involved lately.

"He told me all I talk about is work," she explained. "And he's right."

Lisa's ambition had ratcheted up to high gear. She'd regale Doug with the intricacies of her latest project and how much she'd impressed the clients. She'd surge into a monologue, her voice charged with excitement, as she brought him up to speed on her latest, grand vision to fix the homeless shelter.

"He was feeling totally unimportant," she said. "I knew I had to fix that. The last thing I want is for Doug to feel like he doesn't matter to me."

"So what did you do?" I asked.

"I told him I'd been selfish and would make it up to him," she said, smiling. "I stayed home the next night and cooked us dinner."

Lisa displayed other features typical of people around 5 on the spectrum. She drew inspiration from her grand ideas. She'd become a creative leader in her field, rallying supporters even in the political arena. Her dreams pushed her to achieve and rise above an ordinary life but she never used them to make people feel beneath her. If anything, people felt important in her presence, as if they brought value simply by being who they were. Lisa made the quiet woman in the waiting room light up.

That's a sure sign you're with someone in the middle of the spectrum—they bring out the best in everyone.

Interestingly, they're not an especially modest bunch. They don't need to brag or boast or show off to feel good about themselves, but they're not bashful about their talents either. Lisa, for example, met her husband at a nightclub and she'd approached him. She slipped up beside him, brushed his shoulder, and after a few minute of flirting, invited him onto the dance floor. "Come on," she'd said. "I'm a great dancer—promise."

And she was.

Now you've met people along the whole range of the spectrum, from extreme echoists to extreme narcissists. And you can see that narcissism has many faces, both healthy and unhealthy. No doubt at this point you're wondering: Where do I fall on this spectrum? You may already have some sense just from reading and relating to these stories, but you can get an even better idea by completing the Narcissism Test.

4

—

THE NARCISSISM TEST

HOW NARCISSISTIC ARE YOU?

Before you grab your pen and flip to the test—I know you're itching to—you should know a few things.

First, don't expect to fly through this test. It's not like one of those quizzes you'll find in popular magazines. As you've seen, narcissism is far more complicated than most people think, which means that any test worth its salt is bound to require a little work. It'll be worth the extra effort, however; you'll learn a lot about yourself by the end. You might even be surprised.

Also, this test is not like others designed by psychologists to measure narcissism. Most surveys start with the assumption that *any* narcissism is bad. Answer "True" to "I like looking at my body" or "I am assertive," and your narcissism score begins to grow. Say "True" enough times, and you'll score high enough to be a "narcissist." But there's obviously nothing harmful or destructive about feeling confident about your body or being assertive. And it certainly indicates a lack of healthy narcissism when someone freely admits they're nothing special.

The big failing in present measures of narcissism is their singular focus on the right—mostly the far right—side of the spectrum. What's more, no other test captures the deficits in healthy narcissism, on the left side of the scale. To address these shortfalls, I and my colleagues, Dr. Stuart Quirk, professor of psychology at Central Michigan University and doctoral candidate Shannon Martin, created a new assessment tool called the Narcissism Spectrum Scale (NSS). To ensure its accuracy, we've collected data from several hundred people, young and old, male and female, rich and poor, from all around the world, for a far more representative sampling than the typical college study.

The original NSS consists of 39 questions. To make it easier for people to take on their own, we've narrowed it down to 30 items and simplified the scoring system, which you'll find at the end of the quiz. We call this abbreviated version of the NSS the Narcissism Test. (For more information on the development of the NSS and the preliminary research supporting it, see the references at the end of the book.)

Go ahead now. Get out your pen and get to it. If you're feeling especially brave, and want to get a really clear idea of where you fall on the spectrum, give the measure to a close friend or your partner and ask them to rate you. Other people often see us far more clearly than we see ourselves.

THE NARCISSISM TEST

On a scale of 1 to 5, indicate how much you agree or disagree with each item, using the guide below.

1	2	3	4	5
strongly disagree	disagree	neutral	agree	strongly agree

1. Compliments make me uncomfortable. ____
2. It irritates me when someone gets ahead by being the star. ____
3. I've missed out on opportunities because I was uncomfortable nominating myself (e.g., for promotion or leadership position). ____
4. Sometimes I won't state my ideas because someone else's will be better. ____
5. I often defer to other people's opinions. ____
6. I worry about how other people think and feel about me. ____
7. I'm not sure what I want or need in relationships. ____
8. When people ask me my preferences, I'm often at a loss. ____
9. I blame myself whenever things go badly in a relationship. ____
10. I apologize a lot. ____
11. I'm self-confident, but caring. ____
12. I press on, even when tasks are challenging. ____
13. I take more pride in my achievements when I have to work hard for them. ____
14. I can recognize my limitations without feeling bad about myself. ____

15. I'm happy to acknowledge my faults if it improves a situation. ____

16. I believe both partners contribute to the success or failure of a relationship. ____

17. I can rein myself in when people tell me I'm getting a big head. ____

18. I like to dream big but not at the expense of my relationships. ____

19. I'll take giving over receiving any day. ____

20. Despite setbacks, I believe in myself. ____

21. *I find it easy to manipulate people. ____

22. *I insist on getting the respect that's due me. ____

23. *I expect a great deal from other people. ____

24. *I'll never be satisfied until I get all that I deserve. ____

25. I secretly believe I'm better than most people. ____

26. I get extremely angry when criticized. ____

27. *I get upset when people don't notice how I look in public. ____

28. *I'm apt to show off if I get the chance. ____

29. *I have a strong will to power. ____

30. I'm great at a lot of things compared to most people. ____

* © 1987 American Psychological Association. Adapted with permission from Emmons, R. A. (1987). Narcissism: Theory and measurement. *Journal of Personality and Social Psychology*, 52(1), 11–17. Further reproduction/distribution prohibited without written permission of APA.

Narcissism Deficits (ND):

   Add items 1–10 and enter your score here: ____

Healthy Narcissism (HN):

   Add items 11–20 and enter your score here: ____

Extreme Narcissism (EN):

   Add items 21–30 and enter your score here: ____

UNDERSTANDING YOUR SCORE

The scale breaks down into three "factors." Think of these like three large piles the items fall into mathematically. All three are related to narcissism (or the lack of it). But they predict dramatically different patterns of behavior. Each factor is also a rough indicator of different positions on the spectrum.

As you can tell by the name of each score, the first total represents your placement on the left the spectrum, the second reflects your tendency toward the center (or healthy narcissism), and the third gives you a rough sense of how far you are to the right.

As you can also probably tell, the *only* factor it's good to score high on is healthy narcissism. That's because we designed the scale to mirror the spectrum. It's the extremes (too little and too much narcissism) that cause all the trouble.

Here's a quick guide to what your scores mean.

THE NARCISSISM SPECTRUM

0	1	2	3	4	5	6	7	8	9	10
abstinence		*habit*		*moderation*			*habit*			*addiction*

Do I Have Just the Right Amount of Narcissism?

Is your HN score at least 43? If not, go on to look at the Narcissism Deficits section on page 53.

If you scored 43 or higher and this score is higher than your other scores, skip to the Healthy Narcissism section on page 54 and read all about your blessings. You're right where you want to be, at around 5 on the spectrum.

Do I Have Too Little Narcissism?

Is your ND score higher than 35?

If yes, and it's also higher than your other scores, you're squarely in Echo's range, 1 to 3 on the spectrum. Skip to the Narcissism Deficits section on page 53. You'll find the descriptions fit well.

Do I Have Too Much Narcissism?

Is your EN score 35 or higher? If yes, go to the Extreme Narcissism section on page 55.

A good rule of thumb is: If you score high on Extreme Narcissism, you're at least around a 7 or 8 on the spectrum, regardless of your other scores.

That's because we designed the scale with the assumption that unhealthy narcissists would paint themselves in the best possible light on everything (their usual modus operandi on paper and pencil tests). That means they'd score high on healthy narcissism *and* extreme narcissism.

FINDING YOUR PLACE IN THE SPECTRUM

If you're happy with what you've already learned, you needn't go any further. But to pinpoint your precise location on the spectrum, you'll have to do a little more work. Right now, you may have a rough sense of where you fall. Without more information, it's hard to distinguish between 2 and 3 or 7 and 8. And if you're one of the rare folks who scored, for example, high on EN *and* ND, you'll need to dig a little deeper.

We'll consider each score, in turn.

Narcissism Deficits: Special Fears

The average score for Narcissism Deficits (ND) is 28. If you scored between 28 and 34 (or lower), you're doing fine. But as scores rise, expect to see more problems.

People who score high (35 or above) on ND tend to:

- Suffer from low self-esteem
- Subjugate themselves to their partner's wishes and needs
- Feel undeserving/underentitled
- Struggle to give and receive emotional support
- Feel pessimistic
- Be modest
- Feel anxious, depressed, and emotionally fragile

The two statements that best define this group are "I'm not sure what I want or need in my relationships" and "When people ask me my preferences, I'm often at a loss."

If you scored 35 to 41, you're probably at around 2 on the spectrum.

If you scored 42 or higher, give yourself a 1.

The vast majority of people who score high on this factor *won't* score high on the other two. If you scored high on this one, it's likely this will be your only high score.

If you're at 28 or lower, you're in pretty good shape; you can at least tolerate feeling special enough to benefit you and those you love from time to time. Picture yourself around a 3 on the spectrum. You might even be higher than that. But to find out, we have to examine your healthy narcissism score.

Healthy Narcissism: Special Joys

The average for Healthy Narcissism (HN) is 39. If you scored well below that, 35 and under, keep yourself exactly where you were on the spectrum. A lower score here just confirms that you don't necessarily *enjoy* feeling special (though you might tolerate it).

If you scored at least average, congratulations: place yourself at 4 on the spectrum. You're moving up!

If you scored at least 43 to 46, give yourself a 5.

If you scored 47 or higher, give yourself a 6.

Now comes the good news. People who score higher on this factor tend to:

- Be calm, optimistic, and cheery
- Possess high self-esteem
- Excel at giving and receiving emotional support
- Experience a sense of purpose in life
- Be self-disciplined
- Be trusting, enjoying closeness and emotional intimacy
- Feel deserving, but not overentitled

The two statements that best define healthy narcissism are "I like to dream big, but not at the expense of my relationships" and "I can rein myself in when people tell me I'm getting a big head."

Interestingly, people with a high HN score are more likely to view not just themselves as special (for example, more attractive and intelligent, less selfish or impatient than most people); they also see their partners as better than others, too. They truly see themselves—and the people they love—through rose-colored glasses.

Unless, that is, they rank high on the next score, too.

If you've come this far, you're ready for the big reveal. Have you slipped past the center?

Extreme Narcissism: Clinging to Special

The average EN score is 27. Most people disagree with or feel neutral about the majority of items on this factor.

If you scored 27 or below, you can keep your spectrum estimate where it is. But as people score higher on EN, problems begin to mount.

If you scored 35 to 41, consider yourself a 7 on the spectrum.

If you scored 42 and above, give yourself an 8.

A 9 is entering the territory of pathology. You can't use self-report measures to diagnose a personality disorder. It's not accurate enough—and you need a mental health professional to provide a formal diagnosis. But if you scored higher than 42, you should probably read through the description of what 9 looks like. You're fast approaching it.

High scorers on EN tend to:

- Have fluctuating self-esteem
- Struggle to give and receive emotional support
- Be entitled, manipulative, and approval seeking
- See themselves as better than their partners (and most everyone else)
- Seem argumentative, uncooperative, and selfish
- Seem unemotional (apart from anger and thrill seeking)
- Experience significant conflicts at work

The two statements that most define this group are "I secretly believe I'm better than most people" and "I'll never be satisfied until I get all that I deserve."

People who score high on extreme narcissism *seem* a little

healthier than people with narcissism deficits; for example, they're more optimistic and seem to like their lives and themselves (they choose statements such as "When I look at my life, I am pleased with how things have turned out"). They also claim to be less anxious and depressed than people on the left. But their pattern of apparent strengths pales in comparison to the gifts displayed by people in the center. Their ego is brittle and easily shattered; they protect themselves by boasting about their gifts or blaming others (or even attacking them) when their self-image is threatened. And their relationships clearly suffer from their argumentative, careless approach with other people.

HELP! I DON'T FIT THE NORMAL PATTERNS!

In rare instances, people score high on narcissism deficits and high on extreme narcissism. If your scores reflect that pattern, it likely means that you vacillate between extremes of feeling worthless and feeling superior. Even if you don't say it, you might have impossibly grandiose dreams, usually about being in charge or showing people you're better than they are.

Most people who score high on extreme narcissism would never think to agree with "I often defer to other people's opinions." Likewise, the majority of people with a lot of narcissism deficits feel completely undeserving—the polar opposite of entitled.

But when a self-doubting, withdrawn style combines with a pattern of rage, envy, and extreme entitlement, both scores become elevated. That's the hallmark of *introverted narcissism*. If you show that pattern, you're highly narcissistic, but either introverted by nature or beaten down by a series of failures.

You're at least around a 7 on the spectrum; higher if you

scored 42 or above on EN. This pattern isn't unusual if you feel superior to everyone in your mind, but the world refuses to confirm your opinion. You look like someone who lacks narcissism, but in reality, you probably cling to feeling special—and you're not getting enough attention to feed your habit. People close to you will see the entitlement and arrogance, but your work colleagues probably see someone riddled with anxiety and self-doubt.

By now, you should have a much clearer idea of where you—or people you love—fall in the spectrum. Use that knowledge throughout the book. You'll be better prepared to understand how and why the people around you act as they do. And you'll be able to spot a dangerous narcissist from a mile away.

PART II

ORIGINS

Healthy and Unhealthy Narcissism

5

—

ROOT CAUSES

THE MAKING OF ECHOISTS
AND NARCISSISTS

At a residential treatment center where I once worked as a psychologist I met two patients who couldn't have seemed more different. Jay, a burly, blond sanitation worker, had been admitted after threatening to kill himself on his landlord's doorstep over a rent dispute. Obstreperous and demanding, he'd stride into the common room, barking orders at staff and patients alike. He was so obnoxious that only a few days into his stay, I caught the nurse in charge, a gentle, serene young woman—"the saint," people called her—positively steaming. "Ugh," she exclaimed. "He's the most infuriating narcissist I've ever met!" The other patient, Carol, was a petite brunette with timid eyes who was on disability from her job as a photographer after trying to kill herself with an overdose of pills. Carol followed every rule, never asked for anything, and generally kept to herself. She barely said a word when she joined the community for meals, exercise, and movies. Where Jay grabbed center stage and sucked all the air out of the room, Carol retreated, distancing herself from any interaction.

The odd thing: They were siblings.

What explains why Carol was so far to the left of the narcissism spectrum and Jay so far to the right? The two biggest factors in determining who we grow up to be are nature and nurture, but just how much each contributes is the subject of long and heated debate. When it comes to narcissism, though, nurture holds the trump card. All of us are born with a drive to feel special—it's part of our innate temperament—but whether we tip into the unhealthy wings of the narcissism spectrum to become self-effacing wallflowers or chest-thumping braggarts depends mainly on our environment.

NATURE SETS US UP

We don't arrive on earth as blank slates. We're born with a temperament, that is, a mix of biological tendencies that push us toward either, say, being careful or impulsive, emotionally intense or even tempered, imaginative or practical. Barring severe brain injury that jars our neural circuitry in another direction, we hold fast to our genetic blueprint. For example, a highstrung person may become calmer with age—research suggests we all do—but will never be as placid as someone who's been "an easy child" from day one.

One of the most heavily researched of these biologically based traits is introversion/extroversion. Extroversion drives people to ascend the stage, become the life of the party, and seek new adventures. Introversion makes us a little quieter, less drawn to crowds, and more reflective, taking the time to think before we speak. Whatever form our narcissism takes is bound to be influenced by this inborn tendency; as you've seen, not all narcissists are charismatic go-getters—and a good deal of that has to do with our genetic predispositions.

Narcissism, too, is naturally stronger in some people than others. Signs of a higher narcissistic drive emerge as early as age three—a relentless craving for attention, an outright allergy to rules. Narcissism may also be heightened by lack of another innate tendency that shows up early: empathy, or compassion for others. Much less is known about the basic temperament of echoists, but preliminary findings from the Narcissism Spectrum Scale (NSS) suggest that people on the far left might be born with a more fragile nature—prone, perhaps, to guilt and shame and fear—and therefore more likely to keep a low profile in any environment.

But even the most rambunctious or timid toddlers can become healthy adults with enough love and support. Nature, in other words, may set us up to lean left or right, but we all have the chance to live in the center of the spectrum. What swings us into unhealthy territory—or one version of narcissism versus another—is how we're raised and what culture tells us is important.

NURTURE SLIDES US HIGHER OR LOWER

The key childhood experience that pushes children too high or too low on the spectrum is always the same: *insecure love.*

To settle at the center of the spectrum, children need to feel that no matter what they do—or don't do—they can still count on the people who raise them to listen and offer comfort when they feel sad or lonely or scared. That's the hallmark of secure love, and when children don't receive it, they'll shape their behavior to try to earn love in unhealthy ways, such as chasing after attention (narcissists) or keeping to the shadows (echoists).

Narcissists can be bred in several ways. Parents who seem to notice or celebrate their children *only* when they're standing out—by becoming valedictorians or varsity league players or beauty queens—set them up to scramble after accolades and approval for the rest of their lives. Extroverts with this upbringing are likely to join the group of narcissists who initially seem charming but turn nasty upon closer acquaintance; introverts, on the other hand, may become brittle adults, seething or withdrawing when people don't pay rapt attention to everything they say.

But introverts and extroverts alike are bound to slip higher up the spectrum if their parents constantly intrude and interfere in their lives, grossly misreading—and ignoring—their needs for privacy or space. People who approach parenting this way are inevitably narcissistic themselves, forever prioritizing their craving for control or attention over their children's need for autonomy. Their sons and daughters learn that any time they allow room for anyone else's needs, they completely forfeit their identity. If they're extroverted enough, they'll fight for their freedom, turning up the volume and growing deaf to everyone else, much as their own parents were to them. The solution here seems to be: *If you can't beat 'em, join 'em.*

In contrast, parents who seem chronically emotionally fragile—anxious or angry or depressed—may cause their children to slip a few notches to the left on the spectrum. The children learn that the only way to earn love is to make as little impact as possible on the people around them. *I can't possibly ask my parents for any more—they might cry or scream—but maybe if I ask for very little, they'll love me.* Left-leaning, temperamentally sensitive children, naturally attuned to others, may be at the greatest risk here. A similar result occurs when parents seem happy or fulfilled only when their children praise, flatter, or comfort them. An example might be a mother who needs her child to tell her she is pretty or a good parent. Such "parentified children" learn to echo and mirror their parents' every need or desire, burying their own needs completely.

The type of parenting children receive determines where they land on the spectrum. But for those who wind up on the right, there is another influence that determines what type of narcissist they become: culture. Societies that prize individualism and fame, such as the United States and the United Kingdom, are apt to produce uber-extroverted narcissists who raise navel-gazing to a high art. In contrast, cultures that prize altruism and group harmony, such as Japan and many other Asian nations, tend to create communal narcissists who pride themselves on being the most patient, loyal, and polite people on the planet.

Narcissists and echoists are made, not born. We may begin life with some leanings, but it's the people in our lives and the world around us that influence where we finally end up on the spectrum. Which brings us back to the story of Carol and Jay—a powerful lesson in how the forces of temperament, parenting, and culture can combine and lead to very different outcomes.

Jay, naturally rambunctious, and Carol, timid from birth, grew up in a home ruled by a highly narcissistic father, who flew into a rage whenever his nightly routine—reading the paper and throwing back Jim Beam—was disturbed. Carol learned to take the path of least resistance, tiptoeing around her dad in the hopes that he'd stay calm, while boisterous Jay found the only way he could get noticed was, like his father, to shout and dominate others before they could dominate him. Both siblings displayed unhealthy levels of narcissism, but had learned different coping mechanisms as children that led them to become mentally unstable adults.

Routes to the extremes and center of the spectrum are many. To get the clearest picture of which experiences encourage, or discourage, healthy and unhealthy narcissism, let's meet a few more people at various points along the spectrum.

ECHO: DON'T DREAM BIG

Jean, 62, arrived at my practice late one fall afternoon, after the last of her children had left for college. "Ever since my youngest, Sherry, started her first semester, I've felt a little lost. Maybe it's empty-nest syndrome." She laughed a bit too loudly. Her eyes darted around the room, to the thin purple vases on my fireplace mantel, to the magazines on the table, to me, then to the window on her right. For a moment, I pictured her climbing out the window and dashing down the street.

"Do a lot of people do this?" she asked, picking lint off her pale green slacks "Just talk about themselves?"

"It can take some getting used to," I acknowledged, hoping to put her at ease. She took a sip of water and continued. "It feels weird. People wouldn't dream of doing it in my family." She sat in silence for a moment, gazing at me, as if waiting for my permission to continue. Her eyes, bright green and glistening, kept disappearing beneath short gray bangs.

"Your mother and father didn't talk about their thoughts and feelings?" I asked.

"Definitely not," she answered. "My dad used to say no one really cares what goes on in anyone else's head, so why bore them with it."

"Sounds like a pretty private guy," I observed.

"Not just private," she said, taking a sip from her mug. "He thought it was arrogant to talk about yourself."

In fact, anything that smacked of self-importance seemed to earn silence or subtle disapproval from her family. Her father, Irish American and a devout Catholic, was a stern judge and had often warned her that most people commit crimes out of pride. At night, before bed, he sat puffing on his pipe and rattling off sayings, like "Never get a big head—it's a sure path to trouble." The rules left her feeling hemmed in, quietly ashamed of her own aspirations. Jean couldn't recall a single moment when she felt comfortable sharing her accomplishments.

Her mother had a similar stance. A shy woman who often stood behind her husband at parties hoping not to be noticed, she rarely had much to say to anyone, even to her daughter. But occasionally, she made her feelings quietly clear in ways that left Jean unsettled. Once, as Jean played with a set of dolls, making them leap and twirl in the air, her mother glanced sadly at her as if she were about to cry. "She muttered something about my head being in the clouds." Jean glanced down and then up, peering hesitantly through gray bangs. "She always got quiet when I talked about my dreams of becoming a great dancer."

"That was a dream of yours?" I asked.

"Oh yes!" A smile flickered across her face. "But I knew that could never happen. My parents said they couldn't pay for lessons. By the time I left home, I forgot all about it."

"Do you ever think about dancing now?" I asked

She looked up at me again. I thought she might cry.

"I just need to think about all that I should be grateful for," she said, straightening herself up. "My kids are happy and healthy." I could tell if I pressed the subject she might stop talking again, so I took a different tack.

"Does your husband know about your interest in dance?"

"I doubt it," she said. "He's pretty busy most of the time."

Jean's husband, a sales manager at a retail store, had been mostly absent throughout the week for several months now, but she'd grown used to that. "He never stuck around much when the kids were younger, either," she explained. A decade into their marriage, he started disappearing, a practice he blamed on the travel demands of work. Later she found out he'd had an affair. Had that been hard? I asked.

"I was furious at him, but I got past it," she quickly reassured me. "I focused on the kids—they're the most important part of my life anyway." The way she said it, I could tell the affair hadn't been his first. "But now all the kids are gone, I feel lost. I honestly don't know what to do with myself."

"How can you?" I said. "You learned *never* to focus on yourself—and now, you're the only one left to focus on."

We all need dreams. They lift us up when life becomes hard. They remind us of our potential when we fail and provide freedom when we feel trapped.

Jean's story is a great example of what happens when we're not allowed to dream or enjoy pride. Jean's father and mother subtly turned away from her whenever she scored a small success or dared to imagine herself achieving great things, so much so that she stopped dreaming altogether. Far from feeling entitled, she felt lucky to have anything at all. The thought of expecting or asking for more filled her with shame. Her parents instilled in her a fear that's common to those who live at 3 on the spectrum: *believing yourself special in any way is shameful.*

While Jean's parents were emotionally withholding, other parents may be outright attacking or demeaning. That's what happened to Bill, who lived at 2.

Bill came to see me because he had struggled with depression for years. At 30, he already hated his job as an accountant—a career, he confessed, that he'd never wanted in the first place. So how had he ended up in a profession he despised? Bill loved art; his mother—sadly for him—did not.

"I don't know why you spend so much time doodling and drawing!" Bill's mother admonished him throughout his childhood. When her son showed artistic promise as a teen, she forced him to take math lessons after school. "You need something practical," she explained, "unless you want to end up like your father." Bill's father, a freelance artist who worked from project to project and lived check to check, left the family when Bill was two, an event his mother never tired of recounting to her son.

Bill learned the lesson: an artistic career wasn't just silly, but *destructive*; he'd lost his father who pursued it and he stood to lose his mother, too, if he himself took the same path. In Bill's mind, the only way to gain his mother's—or anyone's— love and approval was to set aside his dreams of artistic greatness. At his mother's urging, he became an accountant. People like Bill often feel guilty for having any needs at all. They *hate* their needs, viewing them as a potentially devastating force in people's lives.

Parents like Bill's, and to some extent Jean's, are often simply mirroring how their own family raised them, discouraging pride and modeling self-effacement. But sometimes they secretly envy their child's talent and accomplishments. Often, they're even mourning the dreams of their own they sacrificed or failed to achieve. They can't tolerate other people's drive to feel special because they, themselves, had to give it up. So as adults they attack anyone who dares to take a dip in Narcissus's pool. When success means rejection by of our loved ones, any dream of greatness feels dangerous. Likewise, highly narcissistic parents leave their children afraid that asking for anything at all—care, love, empathy—might make mom or dad crumble or explode, an experience that's guaranteed to shift the drive to feel special into low gear.

Parents aren't alone in inculcating fear; siblings can do it as well. Threatened by their brother's or sister's success, they become cruel, dismissive, treating the "special child" like a pariah. As envied children become adults, they fear that pride or accomplishment might open them to attack, so they avoid standing out at all costs, often resorting to self-sabotage; they turn papers in late or delay studying for a test. This is the classic example of "fear of success."

Any time an environment punishes or threatens children for striving to be more, they're likely to reach adulthood on the unhealthy left side of the spectrum.

NARCISSUS: DON'T BE ORDINARY

Chad, 27 and single, is a cashier at a specialty food shop. He came to me after his second relationship fell apart. It seems he'd cheated—again.

"I don't know what happened this time," he said, puzzled. "Lots of gay men have open relationships."

"Did your partner know you wanted an open relationship?" I wondered.

"No, but he should have guessed. I told him I don't like to get stuck in one place."

It hardly seemed like an explanation, and even Chad seemed to notice the gap in logic. "I guess I should have been clearer with him. But most of our friends have been in an open relationship. Guess I blew it." He took a deep breath and straightened his back. "Which is all the more reason to be here. My job is all I have left, and I need to keep it." Chad had been short-tempered with shoppers in the past few months and had been warned his job was in jeopardy.

"Do you often lose your temper?" I asked.

He glanced at the clock. His shirt, a black long-sleeve T, was still dusted with snow and he brushed it from his shoulders, frowning. A finger on his right hand carried a bright silver ring with a large ruby.

"I come from a boisterous family," he smirked. "What other people think is loud, we don't."

"They yell a lot?"

"My father sure did," he said dryly. Chad's father, a lawyer, was given to losing his temper with everyone in the family. Chad was his favorite target, but ironically, his favorite child, too. "I had a great childhood," he told me. "My father probably felt stressed from all the hard work, but most of the time he was my biggest cheerleader."

It hardly sounded like a blissful early life to me—his father stomping around every evening, finding fault with everyone: their mother for spending too much money, his sister for dressing "too loosely," Chad for not reading enough. But Chad didn't seem to notice the contradiction. In fact, his fondest memories seemed to be the times his father singled him out. Once, when Chad was nine, he placed him on his knee and said, "Son, you're sure to do amazing things. You've got an incredible mind. Just keep it focused."

"He certainly seems to believe in you," I commented.

"Even when I had a hard time at school, Dad always told me I had greatness in me." Chad picked up one of my pens and started to doodle on a message pad on my desk. "If this job doesn't work out, maybe I can go to law school soon."

The job, apparently, had been Chad's plan to save money. His father had elected Chad to be his successor at the law firm, but he wanted his son to pay a portion of the tuition to law school. "No free rides in life," Chad explained, parroting his father. Chad felt sure he could earn enough to make a sizable contribution, even though he'd been fired from two jobs already because of angry outbursts. He also felt sure he could get into law school, though his academic record was anemic—he'd nearly flunked out of college his last semester. Chad's view of the future was unrelievedly rosy. "I know I can fill my father's shoes."

"Have you told your father about your problems at work?" I asked

"Oh, no!" Chad gasped. "Are you crazy? He'd kill me."

"Have you ever felt like you could talk to your father or mother—or any family member—about something bothering you?"

"Not so much. My mom always says, 'You worry too much, sweetie.' My dad says, 'Great men don't complain. They act.' "

"So you had no one to tell about kids being mean or worries about tests or any kind of distress when you were growing up?"

"Not really," he said, looking a bit sad. "I didn't want to seem weak."

"Sounds hard to me," I noted.

"I guess it could be," he admitted. "I try not to think about it."

"Hence the anger?" I said.

There's a long-standing theory that the unhealthiest kinds of narcissism stem from excessive coddling and praise in childhood. The claim that the self-esteem movement is responsible for the alleged narcissism epidemic reflects one version of this very old notion. The idea is that overly pampered children come to feel, like Narcissus, that they're of divine origin. They can do no wrong. They're smarter, more talented—more beautiful than any other child around them. Keep telling your kids they're special for doing nothing in particular, the logic goes, and they'll grow up to be self-involved, empty human beings. Keep treating them like they're young princes or princesses and they'll start to act like they're royalty. They might even start treating people like servants.

Chad's story, for example, is typical of children who grow up around 7 or 8 on the spectrum. His father certainly lifted him up with praise. His mother did, too, in her own way. We might even conclude, hearing his story, that critics of the self-esteem movement were right: misguided parents and educators drummed it into his head that he was special, sending his life off the rails. But a closer look at his experience reveals something else, all too common in the history of unhealthy narcissists. Praise was about the *only* gift his parents gave him.

For all the cheerleading, for all the reassurance of his gifts and talents Chad heard, the one thing he rarely felt able to count on was empathy and understanding. He never learned that people could be trusted with his feelings and needs. If anything, his family taught him to *hide* his longings for love and caring. The lesson they left him with was that no one can be trusted. The safest way to go through life is to bury your problems. And Chad did so, dutifully. He seemed, in fact, to feel deep shame over ordinary human frailties and failings. And that, it turns out, is the quickest way to engender unhealthy narcissism.

Chad had become so uncomfortable sharing normal worries or fears or sadness that he'd largely given up trying, turning instead to the high that comes from feeling special—that is, smarter, more talented, and sexier than others. He saw, in glimpses, that he'd made mistakes in his relationships or that his anger had become a problem. But instead of seeking comfort or help from others, he soothed himself with fantasies of being a great lawyer or an amazing lover.

Chad rarely felt good about himself without puffing himself

up. Seeking help became difficult for him because the impulse to depend on anyone made him uncomfortable. Any time he looked to someone for real support, he ended up feeling alone and invisible. His father couldn't see Chad at all unless he saw him as his amazing son. So that's the only way Chad could see himself.

This is a common theme in the childhood of unhealthy narcissists. While parents of people on the left discourage pride or dreams or a sense of accomplishment, parents of children on the right often inflate their children's achievements. They need their sons or daughters to be perfect or special or talented—more for their own pleasure than the happiness of their sons or daughters. Parents with their own narcissistic addiction often insist on their children being (or seeming) exceptional in some way. They build their children up in an effort to build themselves up and, in the process, the children come to feel like they're nothing at all if they don't conform to the image their parent has in mind. They're little more than a gleam in the eye of their grandiose parent. As a client of mine once put it, "I feel like a thought someone once had."

Chad lives around 7 or 8, in the range of habit. He seems aware of his problems, even his mistakes, which is typical of people who've grown *dependent* on feeling special. In contrast, narcissists who feel not just uncomfortable, but endangered without their special status make demands out of reflex—and they grow oblivious to the damage they cause in their own lives. This is the range of *addiction*, where Terrel lives.

Terrel, 55, an introverted narcissist who lives at 9, left most of his friends feeling exhausted. He didn't demand accolades; instead, he insisted that people pay unwavering attention to him whenever he uttered a word. It wasn't the normal kind of listening most people expect or even enjoy: his loved ones had to be perfectly attuned to him. If his wife so much as sighed or looked away during a story he was telling, he became angry. "I listened to all your drama with your girlfriends! Why can't you let me finish a story without interruption!" She'd grown weary of his demands, and was threatening to move out.

Terrel complained bitterly that no one understood him. "I'm just different from other people—I see the world for what it is!" By which he meant the world is a hostile place, where he never gets the respect he deserves. Like Chad, Terrel's parents had left him unsure that he mattered in a real (or reliable way). His father, likely a high spectrum narcissist himself, rarely let him get a word out. "Adult speaking," he'd interrupt, trampling over Terrel's stories. His mother, while unpredictable in her support (she fell into strange silences), praised him for being "a sensitive child."

Like many at 9 on the spectrum, Terrel addictively clung to feeling special as an adult because it was the only way he felt like he mattered. It never dawned on him he could get the same high from feeling cared about and understood because he'd never felt that way growing up. He didn't trust anyone to be there for him, so he demanded support and recognition from others as an adult. It was as if he didn't exist at all unless he was the focus of everyone's attention.

HEALTHY NARCISSISM: ENJOY YOUR DREAMS

Enjoying fantasies of greatness without becoming addicted to them requires an ability to feel good about ourselves—to have a solid sense of self-esteem and self-worth, to enjoy attention and praise—but without a relentless need to prove ourselves. People who can accomplish this believe themselves capable of extraordinary things, but aren't devastated when they fall short from time to time. They have a drive to chase after the spotlight but can give up pursuit when the cost becomes too great.

Gina, a 23-year-old art student at a local college, came to see me for career counseling after being encouraged by her parents to seek guidance. "I want a job I'm really going to love. I'd like it to involve lots of people, kind of a brainstorming environment, where people bring their ideas together. That's where I thrive. And I'd love to do some mentoring. I'm not sure how to make all this happen, but I'm here to figure that out."

"You sound pretty certain of what you need."

"Well," she said, with a laugh. "My parents can tell you I've never had trouble telling you what I want in life." After sitting with Gina for a few minutes, I didn't need any convincing.

Brown eyes narrowed, kicking a crossed leg, and flicking her braids, she rattled off the story of her life with the flourish of an accomplished storyteller. What struck me right away was how well her parents had prepared her for this next stage in her life. As an infant, she'd been energetic, even rambunctious at times, and insatiably curious. In grade school, her inquisitiveness had inspired various science experiments with rocks and water. "I had jars full of pebbles all over my dresser for years," she explained.

"How'd your parents feel about that?" I asked.

"They thought it was hilarious. They loved my little schemes. I told them I was going to be the next Madame Curie. My father said, 'Great, we could use one!'" Her eyes twinkled as she spoke about her dreams of being a great scientist, which gave way, eventually, to dreams of being a great artist.

"I suppose it was predictable enough. I loved putting on little shows, with all my drawings. Rocks were my first subject, naturally."

"Your parents sound incredibly supportive. Did you feel like you could go to them when you were upset, too—sad or angry or anxious about something at school?"

"Definitely," she said, barely pausing for breath. "I always knew that if I fell down, my mother would be there to pick me up. I could tell her anything—and often did. I have three sisters, and she still managed to make me feel like the most important person in the world when we were together. My sisters tell me they felt the same way whenever they were with her."

That seemed to be the biggest theme of her childhood. Her parents doted on her. She felt like she mattered, in the deepest sense—that regardless of whether she was successful or not in her various plans, she'd always be special. She often called on the feeling during difficult times.

"When I was seventeen, I entered an art competition. I felt sure I would win, but I got third place. " She paused and smiled weakly. "I know I'm talented. I know I have something important to say—it's just a matter of getting it across. That's what my parents taught me."

"They taught you to believe you could do great things, but you never felt like you had to. Is that right?" I asked.

"I'd say that sums it up," she agreed. And then she went on to tell me about her plans to conquer the art world.

Gina seemed perfectly aware that her intense ambition and super-confident stance were a means to an end. She played on the edge of perfectionism and grandiosity but she never fell off the cliff. No matter what happened in her life, she knew her parents loved her. She trusted that they were there for her, not to rescue her or fix her life when things went wrong, but to understand what she felt. She could share her emotions anytime she needed to because her parents showed her how. They taught her to feel special without her having to insist on it. One of the many lessons she learned over and over was how to fail gracefully.

When Gina turned 13, for example, she experienced heartache for the first time. A boy, one of the more popular in her class, lost interest in her without warning. She was devastated. Her mother promised to sit with her and talk the night of the breakup, but she forgot, too preoccupied with her own problems at her work.

"I was hurt," Gina explained, "but that's not what stayed with me. It was how sad Mom looked. She apologized, profusely, but somehow the only thing that mattered was that she cried while she did it."

After Gina's mother had listened to her for a while, she talked about the problems at work that had distracted her. "It didn't feel like an excuse," Gina explained. "If anything I felt better seeing that even my mother could let me down sometimes and it wasn't the end of the world."

Gina's story is typical for someone in the center of the spectrum. She learned how to accept disappointment without letting it destroy her faith in love. Her mother and father could make mistakes, but they never left Gina feeling like she didn't matter. If a talk went badly, it could be started over. If a harsh word was uttered, it could be taken back and replaced with a kinder one. No one had to get it right the first time. Everyone in the family knew that if they hurt someone, they could always make it up by trying to understand how the other person felt. As a result, unlike people on the left or right of the spectrum, Gina felt secure in the knowledge that if she needed to depend on people for help or care or understanding, they'd be there for her.

Secure love provides protection against many of the world's psychological dangers. It makes people more likely to admit their mistakes and apologize for them, and feel freer to share who they are. They've learned, like Gina, that the people who love them can be trusted to accept them, flaws and all. That's what secure love is: the faith that we can safely *depend* on other people. Parents like Gina's make an effort to understand their child's emotional experience from an early age. They help them name and talk about their feelings, like sadness, anxiety, anger, or fear. They put words to their own emotions, too, just as Gina's mother did. They teach their children how to heal hurt feelings by owning their own mistakes and listening to the pain they've caused. As a result of these many emotional lessons, children learn to give and receive help and love.

This, then, is the recipe for healthy narcissism: A family that encourages (but doesn't require) dreams of greatness and a healthy model for love and closeness. Add them up and you'll find someone who lives near the center of the spectrum. Gina learned to feel important to other people through mutual caring and understanding. It's a lesson that people on the far right or left of the spectrum rarely learn.

6
—

ECHOISM AND NARCISSISM

FROM BAD TO WORSE

You meet a charming, attractive guy named Kyle. He's bright, seemingly taken with your stories, and pursues you. You date for a year or so and things are going well when he suddenly becomes obsessed with a rival at work—"that jerk," as Kyle calls him. He's determined to beat him in the race to the top. Even when you're trying to relax, Kyle has trouble focusing on anything besides the latest battle with his foe. You're beginning to feel alone, even when you're together.

Or you're introduced to a woman, Jesse, who seems like the most caring person you've ever met and yet isn't smothering. She's content to be by herself and just as pleased to be with you. You move in together and go on to live with her for three years. During that time, you notice she always seems to know exactly what you want—but you're never sure what she wants. Even more perplexing, you can't figure out how to comfort her when she occasionally falls into gloomy funks. You want to help, to make her happy, but you're uncertain how. She leaves you feeling baffled a lot of the time.

These are relationships that probably look fine from the outside, but aren't fully comfortable if you're in one of them. Something feels off, but you can't quite put your finger on it. You say to yourself, "He's going through a phase." Or you chalk it up to, "It's just her personality."

And that may be true in many cases. But it could also be that you've fallen in love with a person who dwells in the milder regions of the narcissism spectrum—a *subtle narcissist* or *echoist*.

Unlike the people at the extreme ends of the spectrum, whose behavior is glaringly obvious, subtle narcissists and echoists are much harder to spot. And since you're far more likely to run into them—milder habits are always more common than severe ones—you can easily grow close without realizing there's anything wrong with them at all. Their problems, compared to their extreme peers, aren't nearly as consistent or rigid. Rather, they seem to slip, periodically, in and out of arrogant insensitivity or brooding reticence. They're "with us" for a time, then gone, then back as if nothing happened. Knowing the dynamics that drive this behavior can help you make sense of it.

SUBTLE ECHOISM

Mary, a patient of mine who lived at 3 on the spectrum most of the time, was a 35-year-old barista at a local coffee shop. Her friends called her "the listener"; if people needed to talk through a problem, she was there for them. And if anyone needed help—say, with a move to a new place—she was happy to volunteer her time. Mary never asked for anything in return and she never grumbled about the effort it took.

"I could have been a therapist," she announced proudly at our first meeting.

"How do you mean?" I asked.

"I'd much rather focus on other people's problems. It takes my mind off my own." She yawned, as if the thought of her own difficulties left her bored.

"But that's been hard lately?"

"Impossible." Mary's job was in jeopardy. The landlord, who wanted to sell the building, had been pressuring the coffee shop's owner to move out. Mary hadn't given any thought to moving on—either from her job or in her life. Now she had no choice.

"I have no idea what I want to do," she admitted. "Normally I can handle uncertainty like this but I've been feeling a lot more edgy lately. I feel trapped, anxious."

No one, not even her boyfriend, understood the true extent of the turmoil she felt. She wouldn't tell them. Instead, she went through periods where, to use her words, she "disappeared." She'd lie in bed, depressed, and, when her boyfriend asked what was wrong, she didn't have an answer. And when friends phoned, she let the call go straight to voice mail.

Her boyfriend, upset and frustrated, was now threatening to leave. Mary had been so devastated—she'd pleaded with him to stay—that she finally decided to call me. "This is so unlike me!" she said to me, sobbing. "I don't even know what I want from him. I just want someone nearby."

"What makes it hard for you to talk to your boyfriend and friends about this?" I asked.

"I don't know—I'm afraid it'll drive them away. I don't want them to think I can't handle myself. I don't want them to think I'm needy."

"But it's fine if *they* need you?" I challenged.

"The rules are different," she said, smiling. "I can't explain why."

"Your main rule," I observed, "has been rather singular: try not to need anything at all. As if by giving up expectations and desires of your own—being less demanding—you'll earn people's trust and love. But now you need more, you're afraid your relationships can't handle it. That's why you're in such turmoil. You need some special attention, for once, but you're having trouble asking for it."

Subtle echoists like Mary reflexively focus on other people's needs. It's an unconscious strategy to keep people from rejecting them; in their minds, the less "room" they take up with their own demands and worries, the more likable or lovable they become. People in this range aren't allergic to all attention. Being noticed is fine, as long as they're noticed for what they do for others—being a supportive partner, a productive worker, or an attentive listener. And people like Mary can have wonderful, loving relationships. The only hint that they live in echo territory is the one-sided nature of their support. One always has the feeling with subtle echoists that we need them more than they need us. They much prefer to be in the therapist's chair, not because it helps them feel superior (like communal narcissists), but because it distracts them from their own desires and expectations. They can ask for little things, such as birthday presents and anniversary gifts, or even more attention from a partner, but they keep very close tabs on how much they request, always afraid of crossing the line into selfishness.

For periods of time, they can feel happy and satisfied with their friends and lovers—that is until their needs become harder to contain. Their problems mount when they reach points in their lives where they actually *do* want more. Being the listener or support to others isn't enough. That's when they experience ...

NEED-PANIC

Unlike extreme echoists who constantly clamp down on their needs and desires, subtle echoists simply try to keep their demands to a minimum. Remember, their deepest fear is that asking for too much could drive their loved ones away. The stronger their needs, the more anxious they become about expressing them, and the deeper they go into hiding. As a result, they can slide down the scale to 1 or even 0.

Think of your usually available friend, who suddenly fails to return your sympathetic phone calls after losing her job. Or a warm, supportive partner, who becomes quiet and distant during the holidays, when she misses her parents the most. Or your work buddy who clams up every time you reassure him that his mistakes on a project aren't as bad as he thinks. In these cases, the people may be hiding not from you, but from their sudden need for you.

Alternatively—and paradoxically—subtle echoists can suddenly become clingy and inconsolable. The easiest way to get rid of need, after all, is to get it met immediately, without delay. For people who dread needing anything from anyone, a sudden surge in their desire for support or understanding or even comfort can be frightening, driving them into chaotic efforts to feel better: late-night phone calls, constant texting, or requests to get together more often than ever before. Their guilt and turmoil over their sudden demands can be palpable. Even as they're asking for attention, they wring their hands, as though trying to squeeze the needs out through their fingertips. It's this very pattern that explains why, in our research, echoists were as apt to chase after people as to push them away. Echoists in need-panic are rarely straightforward about what might help them feel better. That's because they've worked so hard to avoid their own needs and expectations that they may not be sure what to ask for.

Once the crisis passes, very often people slide back to their normal place on the spectrum. But if their needs escalate, and they *never* overcome their dread of special attention, they may travel deeper and deeper into echo land over time.

SUBTLE NARCISSISTS

Sherry, 25, a Web designer at an elite firm, came to see me after she and her live-in boyfriend Kevin, 23, began having terrible fights. You wouldn't have known how distressed she felt based on her mood at the start of the appointment. She smiled brightly as she entered my office, complimenting me on the style of the chairs and paintings. "This is the nicest office I've been to so far!" (She'd been shopping around for a therapist.) "I think I found a winner!" Settling quickly into a chair, she plunged into talk. But she gripped her purse tightly in her lap as if someone might snatch it away; it was the first sign that her cheery manner might conceal some worry.

"Kevin just doesn't understand how much pressure I'm under at work. And I don't have time for the drama," she said, nodding as if silently agreeing with herself. "That's really why I'm here." She nervously tugged at the ends of her sleek black hair with her fingers.

Their relationship had been ideal for a time. They met

during college and after graduation lived apart for two years. "Kevin had to finish up at school," she explained. "It just made more sense for him to save money by living on campus."

"When did the problems start?" I asked.

"Four months ago, as soon as I got this job," she grumbled. "He just disappeared on me."

Kevin had lost his mother a year earlier to cancer. Knowing her illness was fatal, he'd been going home every weekend to be with her. Naturally this made it harder for him to spend time with Sherry or his friends. She joined him when she could, but her schedule made it exceedingly difficult. What's worse, after Kevin's mom died, he had to throw himself into graduate school applications—he'd put them off while his mother was ill—and that left him with even less free time. Sherry often ended up going out alone. On top of that, she frequently worked well into the early morning hours. Kevin seemed increasingly lonely and sad. The change had been difficult for them both; neither seemed prepared for Kevin to need much of anything. "He's a pretty self-reliant guy," she observed. "But lately we fight all the time because I'm never home."

Their life in college had been a lot simpler. At the time, Sherry had clearly worshiped the ground he walked on, boasting to friends and family about his charm and warmth: "It's amazing to watch him. He's totally at ease in any crowd. He's like the perfect boyfriend on paper—the right family, cultured, savvy—a born networker." She stopped, grimacing. "Or he was anyway."

"Now you're worried?" I asked.

"Terrified," she said, flashing me a genuinely fearful look for the first time. "He seemed so stable in the beginning. He was my rock."

The more I listened, the more it became clear that most of the fights started because Kevin wanted to talk about something other than Sherry's work. By her own account, as soon as she walked in the door, she started replaying her day, detailing events and conversations that proved her boss hated (or loved) her, and her coworkers secretly envied her. She finished her discourse hours later, declaring that no one, including Kevin, could possibly understand how hard her job was.

"What's it been like for you to see Kevin, your rock, looking so depressed?" I wondered.

Her eyes welled with tears for a moment. "I don't know what to do," she said, sighing. "If I don't knock these designs out of the park, I'll never be taken seriously at my job . . . " She paused, unsure whether or not to finish the sentence. "The thought of screwing this up . . . " She hung her head, cradling it in her hands. "This is my shot." She looked up, crying. "He has to understand."

"Maybe he's too lonely to be your rock right now," I observed.

"I know," she admitted. "But I need this. I'm as smart as anyone in the firm, even the boss. I deserve a shot at running the place." Her jaw clenched with determination. "Shouldn't I be able to count on him to help me prove it?"

Unhealthy narcissism on the right isn't always obnoxiously arrogant or openly condescending. Instead, subtle narcissists are often merely bad listeners, endlessly preoccupied with how they measure up to everyone less. Since winning is an easy way to feel special, they obsess over their numbers at work or compare themselves to anyone who exceeds them in looks or talent or achievement. They're constantly consulting some imaginary scoreboard in their head.

We all do this from time to time, especially when the environment encourages it. Schools, for example, thrive on competitive grading and it's easy to get caught up in our ranking.

But when our score on any metric—whether looks or talent or helpfulness—becomes a *perpetual* concern, we've slipped into unhealthy narcissism. In the 7 to 8 range on the spectrum, extroverted narcissists loudly obsess about standing out, while introverted ones silently tally their position in the race to greatness. But either way, when you're speaking to them, you get the feeling that instead of taking in what you're saying, they're simply waiting for you to stop talking so they can resume their line of thought. They forget, in their preoccupation with their drive to feel special, that narcissism isn't the only—or even the best—way to feel good about themselves.

Sherry's struggle is typical of people who've grown dependent on narcissism for weathering periods of insecurity. In moments, their urge to feel special takes over; they don't lie, steal, cheat, or insult people; but they do become so obsessed with their ranking in the world they can't see the people standing next to them. They can be charming, caring, charismatic, and sensitive for a time. To those close to them, nothing appears wrong. Then, suddenly, the drive to feel special takes over. This is 7 to 8 on the spectrum—the place where Sherry and other *subtle narcissists* live. It's far more common than the brand of unhealthy narcissism found at 9 or 10 on the scale. And that's why it sneaks up on us so easily.

Sherry didn't just want to make a splash at her firm. She wanted to take over the entire business, though she didn't advertise her ambition to her colleagues or Kevin. It lived deep within her—a secret reservoir she drank from whenever she worried life might not turn out the way she'd hoped. And lately, she'd begun to worry quite a bit.

She spent so much time thinking—and often talking—about how she measured up to her peers that she didn't even realize that it was *she* who had left Kevin rather than the other way around. It wasn't her late hours that caused the problems between them but her emotional absence when they were to-gether. She was too preoccupied with the endless dash to prove herself professionally. This is the hallmark of a dependency on feeling special. When trouble hits, people like Sherry, who ap-peared to hover around 7 most of the time, often slide up the scale.

The shift is easy to spot, and it all comes down to one word: *entitlement*. It's the most salient characteristic of the subtle narcissist.

THE ENTITLEMENT SURGE

We all need some entitlement now and then, just like we need to feel special now and then. On birthdays, we feel entitled to a little extra consideration or attention. When we're sick, like-wise, we might feel entitled to more help. Healthy entitlement might even help us say "no" to unreasonable demands and assert ourselves when we're feeling mistreated. But entitlement, at its most extreme, is an unremitting attitude that the world and everyone around us should support our exalted status. It's this kind of entitlement that gives away subtle narcissists.

Entitlement solves a unique problem for the narcissist. Con-vincing ourselves we're better than others requires the presence of other people, and they have a free will of their own. The only way to support a relentless need to tower above other human beings is to bend them to our will—to demand recognition, like a king forcing his subjects to their knees. Extreme entitlement turns everyday interactions into a drug, another chance for a narcissistic high. And the more dependent someone is on feel-ing special, the more their entitlement grows to help them meet their need. Only later, when Sherry needed more support than she ever had before, did her problems become apparent. Kevin *had* to be her rock. Need had become expectation. Sherry felt entitled.

Subtle narcissism is marked by an entitlement surge—those moments when a normally understanding friend or partner or coworker angrily behaves as if the world owes them. It's usually triggered by a sudden fear that their special status has been threatened in some way. Until this point, their need for the world to revolve around them is mostly under wraps, because it hasn't been called into question. Sherry didn't ask for Kevin's support or even try to understand how hard his year had been. In her mind, she deserved his full understanding because she felt so close to her dream of a big promotion.

The entitlement surge of subtle narcissism is a bit like the normally happy drunk suddenly becoming surly and going on a bender, cleaning out the liquor cabinets and storming off to buy more booze. Your usually affable boss suddenly tears into you, worried that the latest project (his idea) is failing. Unbeknownst to you, he's secretly had plans to become the CEO ever since he arrived. Your partner begins complaining about the messy house after your pregnancy, feeling he works hard enough that he *deserves* to come home to a clean house. Your relentlessly supportive friend, who secretly feels no one's as good at helping people as her, becomes cold and bristly after finding out you confided in someone else about your breakup. You always feel a pull from subtle narcissists—a mild sense that you need to support their ego. But after they have an entitlement surge, you feel like all you're doing is boosting them.

For many subtle narcissists, once the crisis has passed, they slide back down the spectrum to less self-involved territory. But the more their fear of depending on people begins to build—if they have repeated breakups, for example—they begin slipping from habit to addiction, convinced that their special status is the only thing in the world that they can truly rely on.

MOVING FROM SUBTLE TO
EXTREME NARCISSISM: ENTITLEMENT
AND EXPLOITATION

If surges don't bring in the needed emotional reinforcement, they can become so frequent that entitlement tips into exploitation. It's the hallmark of the move from dependence to addiction. Escalating entitlement turns out to be one of the key indicators in the difference between healthy and extreme narcissism. In fact, as entitlement peaks and becomes more relentless, people enter the territory of illness, near 9 on the spectrum.

Roger, a 48-year old man who was recently divorced, had violated the restraining order his wife, Susan, had obtained against him. When he showed up at her office and forced her to accept a "letter of explanation," the judge ordered therapy. Roger made an appointment with me, though not, apparently, with the goal of improving himself.

"You think this'll help my custody case?" he asked, squinting at me suspiciously. His hair, streaked with gray, looked like it hadn't been combed in days.

"If you're willing to look at yourself, I'm sure it can't hurt, but understand that isn't up to me." I glanced down at my clipboard, preparing to make a note.

"You gonna pay attention or scribble in your little pad?" he snapped.

"Apologies," I said, a bit shocked by the sudden anger. He stiffened and brushed at his pants. They were as disheveled as his hair and one lens of his glasses—thin black Versaces—had a small crack in the right corner. Roger, formerly a stockbroker, was currently unemployed.

"It sounds like a lot's been happening to you." I ventured

He crossed his arms and glared. "My life's going down the drain," he growled. "I'm out of work and out of money." He reached into his pocket to pull out a cigarette, but then, noticing the No Smoking sign, thought better of it.

Roger had daily panic attacks that left him drenched in sweat. The worst ones prevented him from shopping for groceries. Minutes after leaving his apartment, he'd end up clutching his chest at the side of the road, afraid to get back in his car and drive.

"What happened at work?" I asked.

"Bad investment," he mumbled, sinking in his chair. "It could have happened to anyone." He didn't sound convinced. In the year before their divorce, Roger and his wife had been fighting bitterly over his spending. He'd purchased two Mercedes SUVs and he wanted, despite his wife's objections, to buy some stock he felt sure would jump.

"Susan never accepted my gift for finance," he added. "So I borrowed the money from our retirement without telling her. That was the nail in the coffin for her." His wife had told him early in their marriage that one of the most painful moments of her life was being forced to move out of her home as a child because her father had gambled away all of their money. She'd warned him that if he ever lied to her about money, she'd end the marriage.

"She said I broke her heart, that she could have recovered from an affair more easily than this." He looked panicked for a moment. "She didn't even read my letter of explanation. I was trying to do something big, for both of us."

"You were convinced it would all work out," I observed.

"I deserved the shot, and I took it," he said, grimacing. "No one can take that away from me."

Exploitation is a pattern of doing anything necessary to get ahead or stand out, including hurting other people. Extreme narcissists may suffer incredible withdrawal—periods of anger, sadness, fear, and shame—until they can sneak, demand, borrow, or steal their next dose of attention. If feeling special means taking the credit for someone else's work, so be it. If they have to criticize others mercilessly to feel superior, even if it means throwing their partner's self-esteem under the bus, they will.

Exploitation and entitlement are closely linked. If I truly believe I *deserve* to be treated as the smartest or most beautiful or most caring person in the room, then I'll make it happen. I won't wait for good fortune or goodwill on the part of others to give me what I want; I'll simply take it. Roger didn't wait for permission. He took his family's money because he deserved it. "It's my money," he said, when asked about the decision. "I can burn it if I want to."

Roger also seemed unconcerned about the damage he'd done to his family by investing behind Susan's back. He had a system. He knew he could make it big if only his wife trusted his genius for investing. If he had to lie in the short term, to bypass her worries and get the capital he needed, he had no intention of letting that get in his way. Susan's problem, he believed, was that she lacked faith in his ability.

When entitlement moves into exploitation, other people's needs and feelings begin to matter less and less. Roger dismissed any concern about how devastated his wife would be once she found out about the money. Though he was loath to admit it, his esteem had been shattered by a layoff. He'd grown depressed and anxious, feeling like everyone had left him behind. He'd waited long enough and grown sick of the fights with Susan. "I'm not her father, for Christ's sake," he spat, when she explained the depth of her pain.

For people who approach 9 on the spectrum, the world exists largely for their benefit—and that includes the people in it. When it comes to introverted narcissists at this level, one often gets the feeling of not so much being a person, in their eyes, as an extension of their own body. We exist solely to support their self-esteem and cater to their relentless need to be understood. Extroverted narcissists are apt to leave us feeling less like a person than a lowly, pathetic creature—a bug they reluctantly allow to exist in their presence. Communal narcissists, meanwhile, might leave you feeling like you're the most selfish creature in the room if you can't see how caring or giving they are.

In any case, the toxic blend of entitlement and exploitation (called EE, in the research) leaves people at 9 or 10 so blind to the needs and feeling of others that empathy begins to vanish. Among the "narcissists" on the Narcissistic Personality Inventory (NPI), the people high in EE cause the most damage. Here's where esteem begins to crumble whenever grandiosity fails, where rates of depression and anxiety and even suicidal ideation begin to rise. These are the narcissists who tend to show up in therapy, often vacillating rapidly between nearly delusional fantasies of greatness and devastating episodes of shame. No matter how puffed up they might be at times, their fragility has begun to show. Their puffery feels like the frantic efforts of the Wizard of Oz: vulnerable, frail human beings, hiding behind a bombastic empty show, all in an attempt to distract us from just how small and powerless they feel.

At this level, people become ill, earning them the controversial diagnosis, narcissistic personality disorder. You probably won't meet many people with NPD in your life. But it's important to know what the disorder looks like. Like all personality disorders, it's very difficult to treat. People with NPD need professional help to shift down the spectrum at all—and if they refuse that help, there's very little chance of change. You should think of NPD exactly the way you would any full-blown addiction; recovery's a tough road, but it's impossible when the person denies the problem and refuses help.

As with any mental health disorder, neither you nor anyone else should try to diagnose someone with NPD, even with this book; it takes a trained mental health professional to make such an assessment. For a detailed description of the diagnosis you can refer to the fifth edition of the *Diagnostic and Statistical Manual* (DSM-V), but for now, here's a simple explanation.

As you've learned, we all *need* to feel special now and then. But people with NPD, like Roger, have a strong need, in every area of their life, to be *treated* as if they're special. They're also driven to act special. They're entitled, exploitative, and unempathetic. They tend to be extremely arrogant and condescending, but they can also be shy and full of shame. More often than not, they vacillate between the two stances—feeling special one day and worthless the next.

Either way they demand attention, admiration, and approval or special consideration because they have little sense of who they are apart from how they're viewed by others. And they fight tooth and nail to ensure the impression they make is a "good" one. For the person with NPD, people are simply mirrors, useful only insofar as they reflect back the special view of themselves they so desperately long to see. If that means making other people look bad by comparison—say, by ruining their project at work—so be it. Because life is a constant competition, they're also usually riddled with envy over what other people seem to have. And they'll let you know it.

PSYCHOPATHS: IN THE DANGER ZONE

Extreme narcissistic entitlement, not surprisingly, eventually crowds out not only empathy, but ethics and morals as well; the most coldly unemotional narcissists may also be psychopaths. (Note: not all narcissists are psychopaths, though all psychopaths are narcissists.) Psychopaths have a much lower level of fear or concern or regret than most people; at their most extreme, they seem totally devoid of sadness, anxiety, guilt, or remorse.

Psychopaths' capacity to treat people like means to an end far surpasses that of narcissists' ordinary entitlement. A boastful narcissist might lie, claiming to be a graduate of Harvard when he's really a high school dropout, but it wouldn't dawn on him to steal. A psychopathic narcissist, however, embezzles funds without giving it a second thought if it helps him advance in any way. At their most severe, psychopathic narcissists look very much like the "monsters" Kernberg described; they give little thought to other people, and their rage when confronted with their mistakes, can be terrifying.

They are a true 10 on the narcissism spectrum. People no longer matter. Ordinary human feelings and rules no longer apply. For these narcissists, feeling special becomes the sole reason for existence. They're like heroin addicts who casually kill to get their next dose. Coming back from this level of narcissistic addiction is nearly impossible. And if people do manage to recover, it's only because they learn to seek help— and they stick with it, for many, many years. If you see signs of dangerous narcissism, your best bet, frankly, is to run.

Most of us, with any luck, won't run into criminal narcissists. The damaging people we face are far more ordinary: selfish partners, fair-weather friends, and ruthless coworkers. And as you've seen, it isn't always easy to see these folks coming. Subtle narcissists, after all, are just that—subtle. Until you get to know people—really know them—you may not have the opportunity to see just how offensive, entitled, or manipulative they can get. So how can you spot trouble before damage is done?

PART III

RECOGNIZING AND COPING WITH

UNHEALTHY NARCISSISM

7

WARNING SIGNS

STAYING ALERT FOR NARCISSISTS

You're familiar with the dynamics of echoism now, so you can be on your toes if and when you notice it in yourself or in the people around you. But because narcissists tend to cause us more trouble than echoists—and far more is known about them—I've devoted this section of the book to spotting and coping with them. Luckily, you can use many of the same tactics I present to deal with problems at either end of the spectrum. But there's another reason you need to spend a little more time acquainting yourself with narcissists: it isn't always easy to recognize them, even when they're in the far right end of the spectrum.

Extroverted narcissists in particular are so adept at charming the socks off people that they can easily win us over—at first. They might ascend the corporate ladder, or become the life of the party, or—if you're dating them—lavish you with gifts and attention. Even people with narcissistic personality disorder (NPD) can be great company when they're feeling good about themselves. But as research demonstrates, their charm eventually wears thin—sometimes within weeks, sometimes months or years—and eventually the entitlement and manipulation begin to show.

Are there signs that can alert you early on that you're keeping company with a narcissist?

Yes. One crucial sign: narcissists dodge normal feelings of vulnerability, including sadness, fear, loneliness, and worry. In any relationship, we're bound to make mistakes and hurt others. On a bad day, when our patience is exhausted by problems at work or squabbles with our kids, it's easy to lash out over an innocent question from our spouse like, "Did you pick up the milk?" Or, lost in our own worries, we may neglect to greet our loved ones with a kiss or even say hello. Minor slights like these can be easily repaired if we say we're sorry and acknowledge the hurt we've caused—accidental or intentional—and most people can do so after they calm down. But narcissists often seem incapable of showing contrition or remorse because, as with any kind of vulnerability, connecting with loved ones in this way demands sharing all the feelings that unhealthy narcissism is meant to conceal. And that's precisely what gives narcissists away: they resort to a number of predictable psychological strategies to hide normal human frailties.

The five early warning signs of unhealthy narcissism detailed in this chapter can appear in romantic partners, but they're just as likely to show up in family members, friends, and work colleagues. In fact, they can surface in any relationship from time to time because we're all prone to behaving this way when we're feeling insecure about our abilities, status, or relationships. The difference between narcissists and the rest of us is that people who cling to feeling special use these tactics all the time. And unlike the rest of us, who may turn to one or two of these behaviors during times of turmoil, narcissists often employ them all at once. Even subtle narcissists, who can take years before surging into entitlement, will still give themselves away with their overreliance on these tactics. They're "tells," if you will, hinting at trouble long before more damaging behavior comes along. Study them closely, and you're well on your way to fine-tuning your narcissism radar in every area of your life.

Mark, a man in his early 20s who worked as a bank teller and was about to start graduate school, came to see me because he'd grown troubled about his relationship with his girlfriend, Mia.

"I can't put my finger on why," he said, wrinkling his brow, "but I've been feeling more and more uneasy." He shook his head, confused. "At the beginning, everything was awesome. Mia wanted to spend as much time as possible together. We saw so many great bands on the weekends I could hardly keep up with all their names. She'd just show up with tickets or call me a few hours before a show, and say, 'I've got a new group for us to see.'" He beamed as he talked, savoring the memories. "It was a new adventure every week."

Mia's excitement hadn't been the only thing that turned Mark's head. "She's so beautiful," he exclaimed at our first meeting. He showed me a picture on his phone, the two of them at the beach together: Mia's black hair, thick and lush, tumbled down her back nearly to her waist. "She spends hours on that hair," he said, "getting it to look just right." He placed the phone in his lap. "I won't lie. Her looks got my attention. But that wasn't what clinched it. She also made me feel great about myself."

"How so?"

"For one thing, she called me 'Mr. Right'! She used to lie with me in bed and tell me how lucky she felt to meet the smartest, most handsome guy in the world. That was like the first month we met!" Mark leaned back in his chair, glancing at the ceiling, as if pondering what he'd just said.

"Something troubled you about it?" I asked.

"I guess it felt a little weird sometimes," he confessed. "I mean it didn't seem based on much, since we'd only just started to date."

In retrospect, other things felt off to Mark, too. Mia praised him up and down, but she seemed to lay it on especially thick when he showed even the tiniest signs of liking a band or movie that she did, too. "That's what I love about being with you," she'd coo. "We appreciate all the same things!" Secretly, Mark held back the fact that Mia's favorite band hadn't quite made his hit list.

"What made you decide not to tell her?" I asked

"It didn't feel important. I figured I like their music enough, so why ruin the excitement."

Lately, though, excitement seemed hard to come by. Mia had started arriving later and later for their dates. She had various excuses—she had to fix her hair, she felt tired, she wasn't hungry. Mark found himself just waiting for her more and more often. Once, when she arrived a full two hours later than they'd scheduled, Mia matter-of-factly declared, "I had to finish watching my movie." When Mark glanced at her, dismayed, and began voicing his disappointment, she interrupted, "Don't get all clingy and insecure on me again." She smiled and kissed him, but Mark wasn't reassured by the show of affection.

Mark's insecurity had apparently become a running theme. "If I say anything at all about missing her, or wondering what's happening to us, she tells me, 'Stop worrying about it. We can't spend every minute together.'" Mark scratched his head. "And she's the one who wanted to spend all our time together in the beginning."

"When did things change?"

"Around the time I started applying to graduate school," Mark explained sadly. "She kept pushing me to look at other schools. Ones I'd already ruled out, based on location or lousy research." He frowned. "I remember feeling a hot wave of panic wash over me when she said that. I value her opinion. It worried me to hear her imply I was overreaching with my list of schools." He always left their discussions about the future feeling a little unsettled.

"And Mia?" I asked. "Does she have plans of her own?"

Mia seemed stuck, according to Mark, when it came to her future. She hated her job as a hostess at a high-end restaurant, convinced that her talents were wasted there. "I know I can do something better with my life," she fumed to Mark. "Everyone there is so boring I want to tear my hair out!"

She'd been toying with the idea of attending graduate school, in English or Creative Writing, but she couldn't seem to bring herself to apply. Mark encouraged her to reach out to her father, a literature professor, but Mia refused to. "He could care less," she grumbled. After several fights, he gave up urging her to call home and she hadn't said another word about applying until Mark started his application process. Then, she grew interested again.

"I figure if you can, I can, too!" she told him brightly. But she'd yet to send for any information. When Mark asked her about her plans, she grew irritable. "You're always worrying about something. Can't you just relax?"

"I don't get it," Mark said, collapsing in his chair. "It's like

I've gone from her thinking I walk on water to being a needy mess. She's right, too—I'm always nervous around her now. How can I stop feeling so insecure?"

"Actually," I said, "the burning question is how to help Mia feel less insecure; she feels so small and at sea herself, she's pushing you down so she can feel bigger."

WARNING SIGN: DISPLAYING EMOTION PHOBIA

Human interaction poses a scary problem for narcissists who are, deep down, extraordinarily insecure people. One of their favorite methods of shoring up their self-confidence is to imagine themselves as perfectly self-sufficient and impervious to other people's behavior and feelings. As a result, they don't let on when they feel shaky, or hurt by something you've done or said. Instead, they lash out in anger, which is something we all do when we're upset enough. But narcissists combine it with a show of superiority. They become condescending. They might even point out all the ways you're lacking. Their main goal, in all the bluster, is to hide that you've affected how they feel. Some narcissists won't even admit to their anger, claiming, "I'm not yelling," while they're in the midst of a terrifying tirade. That's how far they'll go to avoid acknowledging emotion.

But emotion phobia can be far quieter than this, too. Because unhealthy narcissism is an attempt to avoid any vulnerable feelings, such as sadness or fear, narcissists often steer clear not just of their own emotions, but also of everyone else's. Mia seemed to fall quiet or change the subject whenever Mark began talking about his fear of graduate school. In part, this is because it reminded Mia of her own insecurities, which she didn't want to share. His sadness touched hers. His fears provoked her own. As soon as he began talking about the future,

Mia's lack of plans invaded her mind. Like many subtle narcissists, rather than talk about her discomfort, she commandeered the conversation, talking excitedly about a new band the two of them could make plans to see.

WARNING SIGN: PLAYING EMOTIONAL HOT POTATO

Whereas emotion phobia signals a deep discomfort with feelings, emotional hot potato is a way of getting rid of those emotions. It's a more insidious form of *projection*, in which people deny their own feelings by claiming they belong to someone else. A friend, for example, might wander up to you, after days of not returning your calls, and ask "Are you upset at me about something?" Given her refusal to respond to your messages, odds are good she's the one who's angry. But instead of recognizing the feelings as her own, she accuses you of harboring a grudge.

In emotional hot potato, however, people don't simply confuse their own feelings with someone else's. They actually coerce you into experiencing the emotions they're trying to ignore in the first place. In this case, a spouse might launch into a rant, laying into you for "being so angry all the time." By the time he's through, you probably *will* feel angry, even if you didn't at the start. That's hot potato. Your partner is getting rid of his anger and whipping it up in you. It's as if he's saying, "I don't want this feeling. Here, you take it."

Insecure about her own future, Mia stirred up worry in Mark. She grilled him about why he'd chosen such hard schools, the implication being that he couldn't get into them. In using this tactic, Mia convinces herself she's got more of a handle on life than Mark does. In other words, she can feel *superior*. Think of the friend who's quick to rate your performance, but damns you with faint praise at the same time ("Not bad—but don't get your hopes up"); the parent who nitpicks whenever you try to make your own decisions ("Why did you do it that way?"); the boss who stares at you in silence whenever you're trying to share your ideas, leaving you tongue-tied in the process. Ever heard the saying, "Don't knock your neighbor's porch light out to make yours shine brighter"? People on the right end of the spectrum love to knock out your lights.

The fact that Mia undermined Mark's self-confidence in the process probably didn't cross her mind for an instant. People at the top third of the spectrum are quick to point out the neediness they see in friends and partners, even when they seem to go out of their way to provoke it.

WARNING SIGN: EXERTING STEALTH CONTROL

Another warning sign is the constant need to remain in charge. Narcissists generally feel uneasy asking for help or making their needs known directly. It confronts them with the reality that they depend on people. For that reason, they often *arrange* events to get what they want. It's a handy way of never having to ask for anything.

Mark often wanted to go hear a new band, but Mia had a list of reasons not to go to a concert—it was too far away, too expensive, too late. But whenever she wanted to see a new artist, the length of the drive or price of the ticket no longer seemed an issue. More often than not, she bought tickets for them ahead of time. In little ways, with each of their interactions, Mia managed to have just the experience she wanted without ever having to ask.

Some other examples: A friend might constantly call at the last minute to cancel every time you've made plans that depart from your usual routine. Such people never say they'd prefer to do something else. They simply preempt the plans you've made. Others might wince or become quiet whenever you suggest something you'd like to do, steering the conversation back to what they have in mind. I've even known people to control their friends by showing up, unannounced, and excitedly convincing them to drop everything and join them in a late-night odyssey. What a fun way to manipulate people! Drag them on an adventure of your choosing.

The effects of subtle narcissistic control are gradual. Slowly, without even realizing it, you fall into the orbit of someone else's preferences and desires, until one day, you wake up and realize that you've altogether forgotten what you might have wanted. It's more like a war of attrition on your will than an outright assault on your freedom. And in the end, the narcissist gets what he or she wants without making a single request.

WARNING SIGN: PLACING PEOPLE ON PEDESTALS

Mia displayed another common habit of unhealthy narcissism—she placed Mark on a pedestal. And in fact, Mark hadn't been the first to enjoy her panegyric, nor would he be the last. Two months into his therapy with me, Mark learned that Mia had been seeing another man—and he, too, seemed to meet her every requirement for the perfect guy.

Why should this be a warning sign of narcissism? For one thing, when people compulsively place their friends, lovers, and bosses on pedestals, it's just another way of feeling special. The logic goes like this: *If someone this special wants me, then I must be pretty special, too.*

In small doses, there's nothing wrong with this. Part of healthy narcissism is the willingness to see our friends and partners as better than they actually are. By elevating the people we care about, we feel lifted up, too. That's why seeing our partner through rose-colored glasses is one of the strongest predictors of relationship happiness.

But there's a difference between looking past people's imperfections and trying to eliminate them altogether. And that's what narcissists like Mia strive to do. They'd rather not even think about all the ways you're an ordinary human being because imperfect people *always* disappoint. So long as Mark could do no wrong in Mia's mind, she'd always feel safe in his hands. That took all the risk out of her depending on him. When you're so afraid of being vulnerable in any way, it's comforting to think you're with a god.

Idol worship always comes with a price, and the clearest one is the absence of deeper connection. Looking up to someone just enough to protect the relationship, giving him or her the benefit of the doubt, allows us to take disappointments in stride and still remain close. But parking someone on a pedestal and insisting he or she quietly stay there ruins intimacy. The space between two people might be vertical, but it's still distance.

Something rang false for Mark, for example, when Mia called him amazing within the first two weeks of seeing each other. Instead of feeling a boost, as if she'd seen what a terrific guy he was, Mark was left with the sense that Mia didn't see him at all. She only saw who she wanted him to be. Mark had to wonder what would happen once she stopped seeing him as perfect. That's the other problem with pedestals. There's only one way off them—down.

WARNING SIGN: FANTASIZING YOU'RE TWINS

Like many people in the right end of the spectrum, Mia seemed to be constantly on the lookout for proof that she and Mark were more alike than different (at the beginning of their relationship, anyway). Indeed, she marshaled evidence in her own mind—and tried to plant ideas in his—that they were pretty much the same person.

It's fun feeling like you've found a soul mate, with all the same passions, fears, ideas, and interests. It's a bit like looking in the mirror. Having a twin provides us with a constant source of validation. With a twin at my side, I can tell myself my ideas make sense, my desires are important, and my needs matter. I don't even need unique talents or beauty to stand out. I can distinguish myself from the masses with a uniquely wonderful relationship. The twin fantasy doesn't demand an illusion of perfection either. We can wallow in—even celebrate—our failings and flaws and still feel great about ourselves.

Narcissists often pair up and wreak havoc under the intoxicating glow of twinship. It's mutually beneficial; even the faintest stars seem to light up the sky when they come in pairs. Perhaps this is why adolescents, struggling with their sense of importance in the world, often buddy up or form groups of nearly identical friends. It helps them feel important in the midst of an adult world that makes them feel insignificant. In a similar manner, young lovers often gaze into each other's eyes, amazed that they've found someone who sees the world just as they do. "We're always on the same wave length," they whisper to one another across the table. Which means they always "get" each other even when no one else seems to care.

Twinning dodges feelings of vulnerability in two ways. First, if you and I are *perfectly alike*—if we're one mind in two bodies—all fear disappears. No difference, no disappointment.

We want the same things. We love, and long to be loved, in exactly the same way. Second, the twin fantasy effectively sidesteps any risk associated with being dependent on someone: since you and I see eye to eye on everything, I never have to worry about you refusing to meet my needs. You just will— and I might not even have to ask. That's exactly how Mia approached intimacy with Mark. She never had any question about her importance. She assumed it, showing up with tickets or calling him at the last minute to get together.

As thrilling as it is, the twin effect can't last. No two people, even identical twins, are ever exactly alike. After a time, when differences become apparent, reality sets in. How people handle this shift says everything about their capacity to move out of unhealthy narcissism. Mia, for example, couldn't stand the fact that Mark no longer struggled in the same way she did. He'd become a bit more confident, clearer about his own desires, and they didn't always match hers. Rather than connect with Mark as a separate human being, and accept or even appreciate their differences, Mia first clung to the twin fantasy, and then, when that couldn't be sustained any longer, she left.

COMMON TRENDS: FAMILY, FRIENDS, COWORKERS, AND BOSSES

Some relationships are inherently closer than others. We generally share more with our friends than our bosses. We also typically reveal more to our family than to colleagues or neighbors. Any one of the warning signs can emerge early on in a relationship, but some require a certain level of emotional intimacy to become fully apparent.

In a family, for example, unhealthy narcissism can easily manifest itself through any one of the warning signs. The twin fantasy is a common ploy in the quietly narcissistic parent. For example, a mother who once dreamt of being an artist might celebrate her seven-year-old daughter's crude attempts at painting while ignoring, and even dismissing, her talent at soccer. Alternatively, a narcissistic sibling might bolster her sense that she's wiser than her younger sister by playing emotional hot potato, questioning the girl's well-reasoned decisions at every turn ("Are you *sure* that's what you want to do?").

Narcissistic friends employ these tactics, too. Your best friend can exert stealth control as adeptly as any date in shooting down your plans for a night out. Likewise, he may steer clear of emotional topics whenever you broach them. But the tactic that appears far more often in friendships than anywhere else is the twin fantasy. This is normal in teens and in adults in their early twenties. But be very careful if you're in your thirties and you feel pressure to be just like your friend. Twinship creates a powerful emotional bond, just short of romantic love—and subtle narcissists often thrive on just that kind of intensity. It's more common in women than men, but male narcissists "twin up" from time to time, too.

Twinship, though rarer on the job, isn't unheard of. Sometimes, supervisors find a sycophantic assistant willing to dress and act like them. Or you might catch a coworker "kissing up" to the boss, placing him on a pedestal. But the most common tactic at work, by far, is hot potato.

Our bosses and coworkers are often looking for ways to feel more competent. What better way to accomplish that than by questioning your every move? Work is all about performance, which provides plenty of opportunity to undermine people's ideas and feelings of competence. Your boss or colleague might ask incessant questions about everything you produce. Or they might suggest an ill-conceived course of action, then blame you when it fails. None of this requires getting to know you, and that makes it even easier to pull off. Like snipers, extreme narcissists often prefer to keep a distance from their target. You'll rarely get close enough to witness their allergy to feelings or hear about their perfect childhood. More often than not, you'll just feel their potshots. But that's also what gives their position away.

Regardless of which signs they display, people who chronically avoid acknowledging feelings scuttle any hopes of deeper intimacy and true, reciprocal relationships. They're too internally preoccupied with their own fears or judgments to accept the gift of genuine sharing.

That's what's so emotionally crippling about unhealthy narcissism. It leaves people so myopically focused on their own sense of importance that they may as well be having an affair with themselves. The only way to reach them, very often, is to clearly and explicitly describe the emotional impact they're having on you. Many people mistakenly think they've done this by admonishing a narcissist or rattling off what they've done wrong. But there are more effective ways to reach them and to effect change.

8

—

CHANGE AND RECOVERY

DEALING WITH LOVERS, FAMILY, AND FRIENDS

Abby, a nursing student in her mid-30s, had been seeing a man named Ned for about six months when she began to spot some troubling signs and came to see me.

"He treated me like a princess at first," she explained, her brow furrowed in confusion. "He called me after each date to chat and hung on my every word! But lately, when we're talking, his eyes glaze over. One minute, I'm telling him about my mother's cancer treatment, and the next, he's bragging about his latest success at work. He just launches into a monologue like I'm not even in the room." She crossed her arms, her face flushed with anger. "Yesterday I couldn't stand it anymore. 'Weren't you listening?' I said. 'My mother is sick!'"

"What happened then?" I asked.

"He said I'm too sensitive—that he *was* listening!" She looked angry again. "I don't know what to do when he gets like that! I see the danger signs you and I talked about, but I either clam up, or we get into a fight when I try to stand up for myself. Is this hopeless? Should we just break up?"

Abby's facing the big question many people grapple with when dealing with a narcissist: When do you know if it's time to go? And, perhaps more importantly, is it *ever* worth staying? The answers depend on whether or not there's any real hope of things getting better.

The problem is we've all had it drummed into our heads that narcissists can't change. They think they're perfect just the way they are, the argument goes, so why should they even try? But unquestioningly accepting this idea backs us into an impossibly tight corner. If all narcissists are irredeemable, then it's madness to want to stay with them. And if we do, heaven help us, choose to stay, we'll try to do the sensible thing, that is, protect ourselves. We'll fall silent or vent our anger, or—like Abby—we'll try a little of each. And none of these reactions will make the relationship any healthier.

When we withdraw, by swallowing our words or walking on eggshells, we only strengthen people's narcissism. In fact, echoists and narcissists often pair up to create a "love" that's toxic to them both. Echoists—especially very nurturing, empathic ones—are drawn to narcissists, precisely because they'd much rather focus on someone else than on themselves. But this approach is sure to create a monster; it confirms the narcissist's belief that being loved means having the *only* voice in the relationship.

When we take the expressive route, unleashing our anger and frustration, we chance fighting narcissism with narcissism. After all, no matter how giving or caring our nature, most of us don't become *more* altruistic when we're under attack. In fact, we usually become less compassionate as we defend ourselves.

In that sense, rage makes narcissists of us all. And when partners spend all their time fighting to assert their own specialness, change becomes impossible.

There is a new more hopeful view, however. Recent studies indicate that the bleak "once a narcissist always a narcissist" view doesn't necessarily hold true. If narcissists are approached in a gentler way, many seem to soften emotionally. When they feel secure love, they become more loving and more committed in return.

In a study cleverly titled "The Metamorphosis of Narcissus," psychologists Eli Finkel, of Northwestern University, and Keith Campbell and Laura Buffardi, of the University of Georgia, decided to put a startling new notion to the test. Decades of research had proven that narcissists tend to devalue love and commitment, cheating on their partners more frequently than nonnarcissists and showing little interest in having warm and caring spouses (they often prefer trophy wives and husbands, partners who validate their importance or attractiveness, for example). But the team wondered: Could the right "reminders" make narcissists more capable of love and commitment?

The researchers recruited 39 female and 37 male undergraduates, who were in relationships of a year and a half on average, measured their level of narcissism, then randomly assigned them to two groups and had them sit at computers that flashed images on the screen. One group was flashed a car, a tree, and a soccer player; the other, a teacher helping a student with homework, a young woman holding a baby, and an older man assisting an elderly woman in a wheelchair. These pictures don't register consciously—they appear on the screen for milliseconds, as quick as a blink—but they do register in the brain and consequently can affect mood and behavior.

What did this subliminal priming reveal? After the flash session, the researchers had people indicate, by tapping "me" or "not me" keys, whether or not they felt *committed*, *devoted*, *faithful*, *loving*, and *loyal* when these words appeared on a computer screen. The number of *me*'s and *not me*'s were then tallied up.

In the group shown neutral pictures, those who scored high in narcissism gave the same response most narcissists do when asked if they're loving or caring or devoted or loyal people: essentially, *not me*. However, in the group shown the nurturing pictures, those who scored high in narcissism tapped *me* on all five traits with startling frequency. In fact, the nurturing-image effect was so powerful that these narcissists' sense of commitment to their partners nearly equaled that of the nonnarcissists in the study. Seeing simple nurturing images, the authors wrote, had *caused* the narcissists to feel more loving and committed.

The team wondered whether this effect would also be true for narcissists in longer-term relationships. To find out, they studied 78 couples who'd been married an average of six years. They scored the participants on narcissism, then had each partner grade their spouse on how well they elicited from them five loving qualities: *nurturing*, *generous*, *friendly*, *charitable*, and *warm*. All participants also rated themselves on how committed they felt to their partners. When the researchers followed up four months later, they found that the narcissists who felt their spouse was adept at drawing out loving qualities considered themselves to be more committed to their partners than they had at the start of the study, sweetly endorsing the statement "I want our marriage to last forever." The change in commitment actually exceeded that of the nonnarcissists. (The narcissists who rated their partners low in these capacities reported no change—their commitment remained low.)

The researchers did a third study, this time with 115 couples who'd just begun living together or who had recently become engaged or married. As always, they first measured the individuals' narcissism. Then, six months later, they asked each couple to discuss for six minutes an important life goal—for example, spending less money and saving more, losing weight and toning up, or finding a more satisfying career. These are emotionally charged topics for even the healthiest of couples. Afterwards, the partners rated the degree to which their mate made them feel "loved and cared about," "capable and effective," or "like a competent person," as well as how much they agreed with the statement, "During the conversation, I felt very committed to our relationship." Narcissists who felt "loved and cared about" by their partner felt more committed than did narcissists whose partner only made them feel more capable and competent. The researchers had now established not only that narcissists can become more committed, but that stroking their ego isn't the way to get them there.

This is all promising research, but the question remains: do narcissists who say they feel more caring in experiments *actually* feel more caring, or are they just giving researchers the "right" answers? To explore this question, psychologists Erica Hepper, of the University of Surrey, and Claire Hart and Constantine Sedikides, of the University of Southampton, performed a series of experiments attempting to enhance empathy in narcissists. In one study, they had narcissists view a video of a domestic-abuse survivor describing her experience and try to relate to what she was feeling ("Imagine what she's going through; try to take her perspective in the video"). Contrary to their reputation for callousness, the narcissists who heard these prompts were moved by the woman's plight. This wasn't playacting, either; they showed a sign of empathy that cannot be faked—their heart rate increased. (Another group of narcissists, told to watch the clip the way they normally would on their own TV, showed no elevation in heart rate.)

More than a dozen studies exploring whether or not narcissists can change have now been conducted—some tracking couples over time, others recording people's emotional responses in a lab—and they all point to the same conclusion: encouraging narcissists to feel more caring and compassionate reduces their narcissism. Thus far, these studies haven't followed participants beyond six months, so we don't know for sure if the changes are lasting. But researchers strongly believe that steadily approaching narcissists in this way can, over time, permanently shift them down the spectrum. How, then, can you encourage this more loving state of mind in a narcissist? How can you respond to them without becoming invisible and losing yourself or lashing out? How can you tell if there's hope or you should leave?

By seeing if your narcissist can come out of hiding.

Always remember that unhealthy narcissism is an attempt to conceal normal human vulnerability, especially painful feelings of insecurity, sadness, fear, loneliness, and shame. If your partner can tolerate sharing and feeling some of these emotions, then there's still hope. But you can only nudge narcissists out of hiding if you're willing to share your own feelings of fragility. As simple as that sounds, it isn't easy. We're all a bit squeamish about revealing our softer side, especially when we feel threatened.

You'll have to dig deep into yourself first. Our most obvious emotions—the surface ones—are rarely the most important. The frustration or anger (or numbness) we feel in the face of a narcissist's arrogance and insensitivity protect us; just below these feelings, however, are the far more potent ones we're usually reluctant to share. We're sad that someone we love has become so hurtful. We're terrified they might leave or betray us. We're ashamed that they've found us lacking (or claim they have). But instead of showing this, we throw on our protective armor. Tears stream down our cheeks, but our voice is full of rage. Or we apologize incessantly, hiding our pain beneath mea culpas, even though, secretly, we feel profoundly hurt. We need to remove this protective armor to give people a chance to understand—and respond—to how we truly feel. It's by doing this that we help narcissists emerge from their emotional bunker and reach for deeper intimacy.

THE THREE STOP SIGNS

These reasons are enough to seek help leaving your relationship

Emotional or physical abuse: insults, name-calling, put-downs, gaslighting (convincing you you're "crazy"), hitting, pushing, or slapping
Psychopathy: A pattern of cold, remorseless lies and manipulation
Denial: refusal to admit to problems and seek help

A few caveats. You have to feel some measure of physical and emotional safety to use the techniques we're about to discuss. If you've seen evidence of outright manipulation—a pattern of remorseless lies and deceit—you might be dealing with a psychopathic narcissist. That doesn't necessarily mean the situation is hopeless—it's still worth being clear with your partner. But it does entail significant emotional risk on your part, and that's where you need to be careful. There's no point in continually opening your heart to someone who's only pretending to change. Some manipulative narcissists are so adept at play-acting and deceit it's hard to know if they're making sincere efforts or just stringing you along.

Which bring us to an obvious point, but one worth making:

The people you love can't change if they're unwilling to ac-knowledge their problems, whether they're alcoholics, com-pulsive gamblers, or extreme narcissists. If they can't push past their denial to some version of, "I think I'm in trouble," then move on. (If this is the case for you, skip forward to "Tackling Barriers to Leaving" on page 127).

Also, it's important to note that none of these strategies is focused on protection—at least not directly. The goal here is to find some capacity for mutual closeness and support. And that demands sharing vulnerability, not stating rules. We'll cover strategies for self-protection on page 125.

EMPATHY PROMPTS

Prompting involves two components: *voicing the importance of your relationship* and *revealing your own feelings*.

Voicing the importance of your relationship generally in-volves making supportive statements, such as *You matter so much to me* or *You're important to me* or *I care about you a great deal*. Declarations like these signal how special someone is to us. They're the kind of reassurance many narcissists don't even realize they miss. They nudge people toward thinking about the relationship, moving the focus from *you and me* to *we*. More importantly, they signal your willingness to offer secure love.

If you can't quite identify your softer feelings, try grabbing a piece of paper and jotting down some thoughts. Take your time—and remember, we never get angry or pull away unless we're in pain. This step is all about describing that pain directly. If you're still feeling angry or emotionally numb, keep digging. The deeper feelings are often ones of loneliness, worthlessness, or inadequacy.

Here are some examples of Mark's prompting in practice:

Mia, you mean the world to me. When you show up hours late for our dates, I feel sad, like I'm not important to you.

Mia, your opinion means everything to me. When you suggest I only apply to easier schools, I'm afraid you don't think much of me.

Here's what you might say to a narcissistic friend in similar situations:

You're my best friend. When you call me selfish, I feel ashamed, like I'm a bad person in your eyes.

I consider you an important friend. That's why I feel so sad when you don't return my calls for weeks.

And here are some prompts you might use with a parent (you can try a variation with siblings):

Mom, you're one of the most important people in my life. So when you question my every move, I feel devastated, like I'm a failure in your eyes.

Dad, you'll always be important in my life. Which is why I feel sad when you give me the silent treatment. It's like I'm losing my father.

When you're prompting, make sure you actually *convey* the softer feeling. Screaming while saying *I'm sad* is rage. It doesn't matter which words you use. If you're not feeling it, don't say it. The feelings have to come out clearly. Take your time with them.

Prompting helps distinguish between people who can change—and those who can't. What you're looking for is genuine empathy, that is, reaching through your own anger (or silence) to share the more tender feelings you've been struggling with. Can your partner, friend, or relative place the relationship—in other words, place *you*—ahead of their coercive attempts to feel special? Can they allow your pain to touch them and say they're sorry or comfort you or just show they understand?

If they can't, you need to view their narcissism exactly as you would any addiction. The "drug" has taken over their lives. Until they're ready to give it up, you might have to give them space; it's likely they have a lot of work to do, and you might not be the best person to do it with. Indeed, it's not your job to be anyone's therapist—only to honestly and clearly convey your own feelings. Once you've done that, you've truly done everything you can. If you've repeatedly tried prompts (say, over several weeks) and seen no softening at all, there's little hope for improvement without professional help.

Failure to respond positively is signaled by people:

Feeling attacked or criticized: "Why are you saying this to me?"

Becoming defensive: "I just got busy, that's all."

Hijacking the conversation: "What about how *I'm* feeling?!"

Blaming: "You're too sensitive."

You're succeeding when people respond by:

Affirming: "You're my best friend, too. I don't want you to feel bad."

Clarifying: "How long have you been feeling sad around me?"

Apologizing: "I'm sorry—I don't want you to feel like a failure."

Validating: "I know my sarcasm hurts you."

If you're in love with someone you suspect is an extreme narcissist, by all means try couples therapy before leaving. You'll find some great options for therapy in the Resources section at the back of the book. Some narcissists can—and do—get better with the right professional help. And if they're resilient enough to acknowledge their problems, you have a fighting chance. But you don't have to stick around for them to finish the work if you're exhausted from trying. Most people, if they're capable of any feelings at all, will melt when they hear empathy prompts. And if they don't, it speaks volumes about their level of addiction. Expect the road to recovery to be long and hard.

Some people tell me, "But I don't feel safe sharing how sad or afraid I am. I'm nervous I'll seem weak, and maybe they'll yell at me." You don't have to approach the relationship in this way if you feel unsafe, but if that's the case, you also need to think about leaving (or with a family member, limiting contact). The research is pretty clear: if we have any chance at creating a healthier relationship at all, we have to take these risks. If you're afraid to try, it might mean the relationship truly isn't safe enough.

My goal in providing this framework is to offer you clarity in your decision. If you share your feelings from a place of vulnerability and the person you care about lashes out or demeans you, consider their response a failure. View it as a sign that they can't—or won't—leave their addiction behind. They're not ready to risk love. But if you do see signs of hope—if the narcissist you know seems to soften when you use empathy prompts, keep using them. They don't just test the capacity for change; they encourage it. If your loved one has any capacity to shift toward the center of the spectrum at all, you'll find out soon enough. And if they don't, you'll know you gave it your best shot.

WHAT TO DO IF YOU'RE THE NARCISSIST

If you scored above 6 on the NSS, you now have a clear target, too. It isn't hopeless, no matter what you might think or what people have told you. But you have to get comfortable with your own feelings and recognize your addiction for what it is: a dodge, an attempt to feel good—even fantastic—without ever taking real emotional risks in your relationships. If you really want to change, follow the same steps I've outlined for your friends and loved ones. Then, take them a step further.

Make a list of the strategies you use to get your special high. Are they arrogance, boasting, or put-downs? How about brooding or rage when you feel "misunderstood"? In the subtler range, do you rely on idol worship or emotional hot potato? These are your protections—each and every one is a vulnerability dodge. If you notice yourself using them, that's your clue: *you're feeling insecure in some way.*

Ask yourself: What's the source of the insecurity? Sadness that your partner doesn't seem to think you're good enough? Fear that your friends might look down on you? The most likely suspects are fear and shame of being unworthy in some way, and sadness and loneliness over being rejected.

Whether you realize it or not, the feelings are there; all evidence suggests they're part of being human (barring severe neurological deficits). So as soon as you catch yourself falling back on old narcissistic habits, take a moment to look for the fear, sadness, or shame lurking underneath. Then take a breath and share these feelings.

Remember: anger and frustration are a cover. If you have even a hint of these feelings when you're speaking to someone who cares about you, you're not taking the risks you need to. The goal here is to test your own capacity to *depend* on people you care about, to move to a place where mutual support and understanding become a way of life. That's what replaces the chronic need to feel special with genuine caring and closeness. You can keep the big dreams or self-assured attitude; just add a healthy dose of empathy and aspire to life at the center of the spectrum. It's the space where you'll not only feel genuinely great about yourself, but also proud of how you treat others.

WHEN CHANGE ISN'T POSSIBLE

No matter how gently we approach them—no matter how hard we work to set aside our own anger or silence—some narcissists won't change. In that case, we might have to leave the relationship. That's a perfectly reasonable choice and one we'll get to in a moment. But in some cases, it's not an option. Cutting off all contact with your narcissistic parent may be difficult. Nor is it feasible to break all ties with a narcissistic ex-spouse if you have children together. You can't ignore your father or the father of your child—not without paying a price. The cost includes overwhelming stress, painful loss, even escalating legal battles. But there's also no safety in continuing to open your heart to someone who's careless with it. So what then?

This is the realm of management, not change. Self-protection should become your primary goal. Limit contact if you can, just like you would with any toxic relationship. But you might also benefit from a few simple strategies and rules. We will discuss some of these strategies in more detail in Chapter 9. But for now, keep in mind that the goal is to manage the narcissism, not foster closeness. In the meantime, you might try using a connection contract.

In a connection contract, you state clearly and simply what has to happen if the person wants you present. It's a way of setting limits by providing rules and expectations.

This is the way a son could explain a connection contract to his mother:

I'm not comfortable with yelling and criticism. If I hear either, I'll leave. I'd like to see you, but it's up to you whether or not I'm able to stay in the house as planned.

A mother, explaining a contract to her ex-husband and co-parent, might say this:

We need to stay focused on custody planning for the holidays. I'm happy to have that conversation tomorrow, but if I hear accusations, blame, or other attacks, I'll take that to mean you're not able to have the conversation and we'll have to come back to it later.

A woman explaining a contract to her housemate:

We need to talk about the cleaning situation and how to set up a schedule. If the talk becomes another laundry list of my problems, that will show me you're not ready to make the schedule yet and we'll have to set it aside and take it up again at another time.

The goal of a connection contract is to explain which behaviors will end the conversation. The emphasis is on what keeps you present, not what makes you happy. If you've reached this stage, your presence is as much as you should promise.

In contrast, we have a choice with dates and partners. We don't have to stick around. Unfortunately, even when we've decided leaving is the best option, it can still feel downright impossible.

There are powerful emotional barriers to saying goodbye. And you need to know how to handle those, too.

TACKLING BARRIERS TO LEAVING

Anna, 32, divorced, had been seeing me for four months to explore whether or not she should continue dating her boyfriend of two years, Neil, a man she described as "infuriatingly self-centered." She'd done a fantastic job of prompting Neil when he became condescending or dismissive.

"I know he has it in him to be caring," she said, sniffling. She unbuckled her purse and pulled out her phone, placing it in her lap, then dabbed her eyes. "I told him I really loved him, but I felt utterly worthless when he raised his voice or criticized me," she said, tapping at the screen. "And this is the message he sent me afterwards: 'Anna, I think we're just different people. You're just more fragile than I am. I don't blame you for it. It's just how things are.'"

She brushed away a few strands of her thick brown hair and glanced down at her pendant, a tiny gold angel—one of the first gifts Neil had given her. "I don't think I can do this anymore."

I understood Anna's hopelessness. She'd tried, valiantly, over the past few months, to tell Neil how sad and nervous she felt when he became argumentative or lectured her. Regardless of how gently she approached him, Neil seemed either unable or unwilling to open up about his own insecurities, instead of tossing them at her or hiding them with bravado. She knew he worried about his career, for example, especially with the sweeping layoffs in his field of investment banking. But he refused to talk about his concerns directly.

"I tried what you said. I even told him if he felt worried about his success at work, I'd be there for him, no matter what," Anna added.

"What happened then?"

"He waved his hands and said 'I don't need sympathy. I just need a smarter boss who can see what I bring to the table.'" Anna's voice became quiet. She was cradling the pendant again. "I know what I have to do. I have to end it with him, and I'm trying. But I keep asking myself, what if I haven't been kind enough? I can lose my temper, especially when he talks to me like a stubborn toddler." She shuddered, her anger rising up.

"It makes sense you'd be angry," I said. "You *have* to protect yourself some way, and if he becomes condescending or dismissive when you try to be open about your sadness or fear, he leaves you little choice."

"Maybe I just need to stop feeling so resentful."

"Or maybe," I said, "the only way you can feel any hope now is by blaming yourself. If Neil *can't* change despite your efforts, then it truly is over. That's not an easy truth to face. Maybe it's easier to tell yourself you're the problem than to accept the possibility he won't change."

ESCAPING SELF-BLAME

The more time we spend with people, the more they literally become a part of us. We think of ourselves, often, not just as individuals, but one point in a vast network of human connections: I'm not simply Craig or Dr. Malkin, but Anna's therapist, Jennifer's husband, Eugene's son. Our identity is tied to the people we love. When these strands of connection become stretched or frayed by anger and pain, we fight to hold on, partly because we're fighting to preserve a part of ourselves. What began as a bond soon becomes a tether.

When Anna decides she can't see Neil anymore, she ceases to be "Neil's girlfriend." It's one loss in a series of many. They'll stop living together. They'll stop eating together. They'll have to sort through the evidence of their identity as a couple—the furniture and assorted knick-knacks—and decide who gets what. The process of extracting shared possessions from "Neil and Anna, the couple" and redistributing them to two, separate people can be as painful as a root canal. Which is why instead of leaving, we often find reasons to stay. One of the more insidious ways we do that is through self-blame.

Self-blame comes in handy when a relationship no longer works and leaving feels too painful. If we convince ourselves that someone's being hurtful or insensitive because of our own failings, there's still hope. All we have to do is improve. If *I'm* the problem, then the happiness of the relationship is entirely in my hands. It's a solution that preserves hope at the expense of our self-esteem.

This is the trade Anna had made in her own childhood. Her father, who drank heavily, often exploded with rage. Rather than accept her powerlessness in the face of this, she decided somewhere along the line that if she just became more obedient or thoughtful, he'd become nicer. With Neil, she'd continued to find hope the same way—and now, it kept her trapped. Regardless of how hard she worked to leave, her self-blame pulled her back.

One way you can liberate yourself from this kind of self-criticism is by confronting a feeling you've probably come to fear more than you realize: disappointment.

Chronic self-blamers bury their disappointment because, in

the past, voicing it might have made things worse. For many people, the fallout from daring to tell their family *That hurt my feelings* or *I really wanted you at my recital* would have been too great. In Anna's family, even when she didn't feel hurt, angry yelling and deafening silence had been the norm. Her father made her feel like a burden when she so much as hinted at feeling unhappy, shouting or sulking until she fell silent. Either way, it became easier for her to swallow her disappointment by taking his message to heart: *You're the problem. You expect too much*.

Remind yourself: You have a right to your disappointment. If you share your needs and feelings and it actually drives the person away, then you can't be happy in the relationship. The solution isn't to slide down the spectrum and become Echo. Recognize self-blame for what it is: a powerful fear that you'll lose love if you ask for what you want. It keeps you stuck in the wrong relationship, with someone who needs you to bury your needs. The only way to find out if they can give you more care or attention or empathy is it to invite them to. And you can't do that if you blame yourself for what's missing.

Disappointment, far from being a threat to intimacy, often deepens it. Being clear about when your relationship leaves you feeling neglected, alone, unworthy, or small puts you back in touch with your own needs. It brings you closer to your lovers and friends. It teaches them how to love you. And there are some simple steps to getting back in touch with healthy disappointment:

- *Create healthy boundaries.* If something hurts, say so. It's not your job to protect your partner or friend from knowing they've behaved badly. By all means, share the disappointment in a vulnerable way. It's your best chance of being heard. But don't let people think you're happy when you're not. That's Echo's trick. If they can't tolerate hearing that you're hurt, they're likely to go on hurting you.

- *Check your self-blame at the door.* When something upsetting happens between you and your partner or friend, keep in mind your fear of losing them is bound to throw you back into self-blame. Instead of asking *What have I done wrong?*, ask *Am I feeling disappointed? Am I afraid to say something's wrong?*

- *Don't confuse empathy with responsibility.* It's fine to try understanding why someone's feeling upset, even when they've hurt you. Maybe your last few comments came across as cold or critical. But you can always correct that by offering a sincere apology. It's your partner's choice to handle their upset by lashing out. Don't make yourself responsible for anyone's actions but your own. That's just another way of blaming yourself instead of feeling disappointed.

ESCAPING THE EXCITEMENT TRAP

Months later, Anna ran into another problem people often face when they end a relationship with a narcissist: boredom.

"I'm enjoying my time with my new boyfriend," she explained, wincing, "Tod's sweet, and charming—a cute guy. But he doesn't stir me up the way Neil did."

"How do you mean?" I asked.

"Neil seemed so self-assured, especially in bed. Sex always felt like fireworks." She smiled, lost in the memory. "Don't get me wrong. I'd never go back now. But I keep hoping I find someone with the same chemistry we had. Can't I get excited with guys who aren't bad boys?"

The answer is yes. But first, Anna has to understand what made her so excited with the wrong guys.

Faced with Anna's situation, many people simply conclude that they're mysteriously and uncontrollably drawn to the wrong people. There are nice guys, on the one hand, like Tod, who offer security and stability and the hope of lasting love, and bad boys, on the other, who offer so much excitement and intensity that it's almost worth putting up with them. Many of these bad boys, like Neil, live on the far right of the spectrum. And it's not just women who struggle with this dilemma. Though it gets far less attention, men have their own version—the "bad girl" phenomenon. One of my clients, Jeff, once complained to me, "Why are all the crazy women so sexy?" It's a puzzle that makes a lot more sense once you realize that our feelings of attraction and excitement often intensify when love feels the *least* certain.

Romantic uncertainty often turns us on. It stirs up feelings like fear, anger, and jealousy, all of which enhance attraction through something psychologists call *arousal*. This isn't the same thing as sexual arousal. Think of it more like a jolt of energy that accompanies any intense feeling and courses through your nervous system. A big dose of arousal ramps up our feelings of attraction. Anxiety excites. Anger entices. Terror titillates. Unfortunately, as far as our bodies are concerned, uncertainty is as good a source of passion as any other feeling. That leaves us at the mercy of narcissists like Neil, who are all too happy to bring us excitement in the form of a constant roller coaster ride of will-he-call-or-won't-he?

To make matters worse, we're often our own worst enemy when it comes to finding safer excitement with more loving partners. We make secure relationships boring.

Freud, true to fashion, didn't miss the pervasiveness of this problem: "Where such men love, they have no desire, and where they desire, they cannot love," he wrote, describing male patients who played out their deepest desires with women

they felt the least commitment to. Their most intense fantasies flourished in the emptiest relationships, with prostitutes
or mistresses. Will our partners still accept us if we reveal our
hidden and wildest desires? Or do we have to present a sani-
tized version of ourselves—safe, reliable, willing to rein in selfish lust? It's a conflict that wreaks havoc with our love lives,
forcing people to find the most intense passion, not in loving
relationships, but in affairs and pornography. We can't escape
the excitement trap of bad boys and girls until we start taking
more risks with the people we love. That starts by putting ourselves in charge of our own excitement. There are a number of
ways to do this:

- *Open up.* Be more direct about your needs and feelings.
  Use empathy prompts. Not only is this crucial in developing secure intimacy, it also ramps up the excitement when
  you're dating. Nothing's more arousing than sharing all of
  who you are and feeling accepted. Being honest about what
  we want and need always entails risk and, because uncertainty is inherently arousing, it builds the excitement. It's not
  the passive, panicky brand we feel with people like Neil. It's
  something far more powerful: *secure passion.*
- *Own your desires.* Sex isn't about purity. It's about imagination and freedom. It's about acting on desire as it emerges—a
  truth bad boys and girls seem to get. In contrast, many of us
  become so concerned about the feelings of the people we love
  that we tie our desire in a knot.

Before Neil, Anna's sex life had been relatively reserved. She
enjoyed sex, but she never felt free. In contrast, Neil, like many
outgoing narcissists, didn't worry about what Anna thought of
him. If something turned him on, he'd try it. He never coerced
her, but he did lead her on some amazing sexual adventures.

His narcissism, expressed in the confidence of his moves, gave Anna permission to act in ways she'd never dreamed of in her marriage. But like anyone who fears untamed sexuality has no place with someone they love, Anna relied on Neil to bring it out. The allure of bad boys and girls lies partly in the room they provide us to be dirty while still believing we're pure.

It's not me, we can tell ourselves secretly. *I can't help myself. He's wild. She's trouble. I'm never like this. I never do this.* And yet, here we are, doing it. We chase after bad boys and girls, in the end, to reclaim our own abandoned desires.

I encouraged Anna to experiment with Tod. A few days later, she sent her own sexy text messages (something she hadn't done with anyone but Neil). She also initiated sex more often. And slowly, as she created her own sense of risk and adventure, she reclaimed her previously disowned desires. To her delight, Tod opened up more, too, and she found herself more happily looking forward to seeing him.

Ask yourself, *What did I do with my ex that I'm not doing now?* Are there experiences you had while pursuing someone, like being seductive or flirting more, that you're not doing when there's less need to chase? Did your ex introduce you to fantasies or sexual experiences that you enjoyed but feel reluctant to enact? Write them down. Enjoy them. Recognize them as your desires, too.

- *Experiment with arousal.* Remember that any intense feeling can enhance attraction. Novelty—when we expose ourselves to new experiences—is a proven aphrodisiac. New experiences trigger the release of dopamine, a brain chemical associated with excitement and reward. Dopamine keeps us coming back for more, whether the excitement we crave is a person or a drug. Our partner becomes exciting by association. Narcissists often drag people into adventures (and drama) that get the dopamine flowing (remember Mia?); learn to generate some of your own. Challenge yourself to try out that new restaurant with your dates or bring them along for dance lessons. Bring a little adventure to your dates with the nice guy or gal. It's an easy way to create secure passion.

ENDING FRIENDSHIPS

Much of what applies to ending a romance also applies to ending a friendship. Self-blame can be a huge barrier to seeing the limits of a friendship, whether the problem's narcissism or any other block to healthy, mutual dependency. It's often far easier to explain away our friends' actions by telling ourselves we're being too touchy or judgmental. In some ways, the fact that we expect less from friends than we do from partners makes things even more confusing. We're not committed to them for life. Our spirit isn't nearly as crushed when they blow us off. When does narcissism become enough of a problem that it's not worth investing in the friendship anymore?

You have two choices with a friend: you either accept the relationship for what it is, limitations and all, or you end it. The first option means lowering your expectations. You've accepted that you can't truly count on your friend—he or she's a work-out buddy or someone you share a drink with. These kinds of friendships can be fun in a limited way, but you have to ask yourself what they add to your life. And be honest. If you have to find other people to turn to in times of need because they're more reliable or understanding than your friend, who are you staying for—you or your friend?

Excitement, too, can be a problem with friendships. Even though you don't feel the same longing for a friend as you do for a lover, you might miss all the fun times. Outgoing narcissists really can be incredibly witty and exciting at their best. It's easy to miss them. But remember the effects of arousal apply to friendships, too. Your narcissistic friend might be fun by association. You look past the drama because you wind up at amazing new clubs or parties. There's no need to give any of that up, but you might need to get better at providing adventures for yourself. The search for new sources of arousal also helps you fend off cravings to hang out with your old pal. You can keep the fun *and* focus your energy on better friends.

It's always a painful experience realizing that your partner's or friend's unhealthy narcissism prevents them from seeing just how much their behavior hurts you. But you can manage the fallout a lot more easily if you avoid blaming yourself and falling mindlessly back into old habits. At least then you'll be clearer about when to stay and when to go.

You now know what do when narcissism threatens your capacity to love and feel loved. But our lovers and friends aren't the only people who shape our lives. Freud once wrote that mental health is the ability to love *and* to work. What happens when you encounter narcissism on the job?

9

—

COPING AND THRIVING

DEALING WITH COLLEAGUES AND BOSSES

Jane, 43, a designer at a small software company, had called in sick more frequently in the past year than she had during her entire nine years with her company.

"It's this new project manager, Drew," she explained to me. "I feel nothing but dread whenever I imagine having to see him. I can't keep doing this. I'm burning through all my sick days."

Drew had been recruited for Jane's company because of his successful record launching new products. Jane's CEO had been particularly impressed by Drew's rescue of a troubled project—a new networking application—from the brink of disaster with a clever design breakthrough. He was strong at innovative design, but even stronger at alienating coworkers.

"The execs put up with him," Jane continued, "because he treats them like kings and queens." She reached for her coffee, blowing on the lip of the cup. Her hand trembled as she took a sip. "I'm a wreck thanks to this guy. He automatically dismisses every idea I bring to the team. Half the time I just sit in silence."

The rest of the team felt useless in his presence, said Jane. Instead of listening to their ideas, he generally grilled them about how their ideas could make the company money. "He gets this look in his eyes when other people are talking. It's like he's bored or unimpressed. Then he starts tearing into them. 'That doesn't pop the way it should,' he'll say, or 'I tried that already on another project and it didn't work.' Nothing's ever any good unless it comes from him."

"Have you tried talking to him?" I asked

"There is no talking to this guy," she said, sighing. "Any time anyone approaches him with a complaint, he walks away. The higher-ups are in the dark. He's a total yes-man with them, so they have no idea what's going on."

Jane had been suffering. In the mornings, before work, her stomach churned; at night, she'd lie awake, replaying the worst moments. She'd been especially haunted by one incident. In front of the whole team, Drew had insulted her ad layout and color scheme as "dull and unimaginative." Yet a week later, he'd taken her work to a meeting of higher-ups and claimed credit for it. No matter how hard she tried to put memories like this aside, they came rushing back and, just as she began drifting to sleep, her mind would start to spin again, churning out fantasies: in one, she'd stand up to Drew with a snappy, sardonic comeback; in another she'd put him in his place, with a blistering speech about his killing team morale.

"But I don't do any of that. Instead, I just keep trying to avoid him."

"Does it work?"

"Not really. He criticizes me for not talking to him. He says I'm not keeping him in the loop." She sipped her coffee again, her hand trembling even more than before. "I'm losing hope. Isn't there anything I can do besides quit?"

"Absolutely," I said. "Let's start by finding out if he can treat you a little better. But he can't do that if you don't hold him accountable. That doesn't mean dressing him down in front of the entire team. But it does mean encouraging him to treat you better."

It's no coincidence that Jane had seen her doctor four times over the past year for one illness or another—the flu, a sore throat, back pain, neck pain—nor was it surprising that her body took longer and longer to recover. The stress had clearly been taking its toll on her immune system. She probably wasn't the only employee suffering, either. Managers like Drew literally make people sick.

Increased sick days exact an immediate and obvious price. Everyone either falls behind or has to work twice as hard to compensate for absent colleagues. Bad employees cost companies, too, in lost production and, eventually, in loss of reputation as a good place to work. What explains disrespect and bullying in the workplace? Often, it's extreme narcissism.

Unfortunately, disturbingly little research exists on how to manage all this damage. In one of the few studies on coping with narcissistic coworkers, people reported using five tactics most frequently: ignoring, confronting, befriending, quitting, and informing management. Those who relied on informing management or quitting seemed pleased with how things turned out, but the same couldn't be said for those who tried ignoring, confronting, or befriending. These strategies, they reported, got them nowhere.

In fact, ignoring the behavior—Jane's approach—often creates more problems than it solves. Keeping a low profile can make narcissistic coworkers or bosses *more* worried about their own performance. That's bad news for everyone. The more anxious people like Drew become about failing at their job, the more arrogant and insulting they're apt to become. Be as cautious about becoming Echo in the workplace as you would with a partner or friend.

Confrontation doesn't seem to improve these situations either. Criticizing an extreme narcissist's manner ("Stop interrupting people!") or pointing out their mistakes ("That slide's completely wrong") usually makes things worse. They can't absorb honest, accurate feedback; instead they become more angry and aggressive, and the already mistreated worker is bound to face a verbal lashing. Besides, where power differences exist, as they did between Jane and Drew, such frank feedback might not even be possible. Few people feel comfortable correcting their boss, let alone an insufferably conceited supervisor or CEO.

So how can Jane—or any of us—stand up to people like Drew without making the situation worse?

The lesson from research is that people only slide down the spectrum when they're reminded of the importance of their relationships. Change doesn't come from telling them off for being too success-driven, ruthless, or manipulative; it comes by showing them the benefits of collaboration and understanding. You've already seen, for example, how a supportive, caring approach can over time deepen a milder narcissist's commitment to his or her partner, and being instructed to imagine someone's pain actually increases their empathy.

Many of these studies examined couples over time. But now it appears that even outside of close relationships, narcissists can be nudged toward greater caring and compassion—and the empathy study is but one example. In another experiment, researchers had narcissists read a passage filled with words like *we*, *our*, and *us* and count the number of pronouns. This simple activity not only made them more willing to help people in need (by giving them the spare change in their pockets, for example), it also made them less obsessed with becoming famous! It's as if the mere reference to relationships reawakens a part of a narcissist's brain—the one devoted to caring and consideration, rather than to fame and fortune.

What we need, then, are strategies designed to remind narcissists of their place in the world of people. To reactivate their blocked empathy—and light up the areas of their brain devoted to concern and consideration. Most people toward the far end of the spectrum aren't accustomed to thinking about other people at all; getting ahead, for them, has become more important than getting along because they've never experienced much success with creating trusting, close relationships. But they're much more likely to try again if they think that "getting along" can also *help* them "get ahead."

In other words, reminding narcissists of the benefits of mutual respect and caring, while supporting them in reaching their goals, might be the easiest way to slide them down the spectrum. Dealing with unhealthy narcissism in the workplace is a two-part strategy involving, first, *protecting yourself*, by creating boundaries and making requests, and second, *nudging the narcissist*, by highlighting the benefits of consideration, collaboration, and respect.

In the following pages, you'll find six interventions. The first three are self-protective; the final three are nudges you can employ to try and slide narcissists down the scale. Each one's a discrete strategy, but you can (and should) try to combine them whenever possible. The best approach involves protecting yourself *and* nudging the person you're dealing with.

But first, a few caveats: All these techniques are based on research with narcissists in the habit range. They may not work with people who reach the level of full-blown narcissistic personality disorder (NPD). Severely disordered bosses and co-workers aren't likely to retain their employment for long, but if you're unlucky enough to run into one before he or she bottoms out, you should still test out these strategies, especially the self-protective ones. Think of them as a way of assessing the hope for change, like prompting in love relationships.

Keep in mind, however, that where bad behaviors cross the line and become abuse, which certainly happened with Drew at times (he called Jane "incompetent" after one meeting), the organization, not the individual, bears responsibility for the problem. *Bullying demands systemic and legal intervention, and it's not your responsibility to stop it; it's your employer's.* You'll find specific help in approaching management in this chapter under "Going Higher" on page 155.

Not all narcissists are bullies, but many are, and you need to recognize them when you see them—a task that's often easier said than done. Bullies sometimes sneak up on us, the way a mugger ambushes people with their guard down. Most of us don't *expect* outright rudeness or dishonesty at work, so we tend not to believe egregious behavior when we encounter it. A sort of reflexive denial sets in: *He didn't really just call me an idiot. She really didn't just blame me for her error.* We convince ourselves it didn't really happen.

Alternatively, when we do recognize we're being slammed, we have a tendency to blame ourselves, just as we would in an unhappy love relationship. *I'm being overly sensitive. I just need to work harder. I really screwed up.* We explain away our colleague's bad behavior for the same reason we dismiss the cruelty of a parent or lover. We need them, we need the job—to eat, to pay the rent or mortgage, to provide for our families. We can't simply quit, any more than a child can just waltz out the door of an unhappy home.

So we deny or dismiss, and hope tomorrow will be better.

Sometimes, we're right; it was just a bad day or week. But sometimes, the put-downs or finger-pointing increase over time. They become a pattern. According to research by psychologists Gary and Ruth Namie, of the Workplace Bullying Institute, the most common bullying behaviors are:

- Blaming mistakes on other people
- Making unreasonable job demands
- Criticizing a worker's ability
- Inconsistently applying company rules, especially punitive measures
- Implying a worker's job is on the line or making outright threats to fire
- Hurling insults and put-downs
- Discounting or denying a worker's accomplishments
- Excluding or "icing out" a worker
- Yelling, screaming
- Stealing credit for ideas or work

Any one of these might show up on the job from time to time. But if you see many of them, frequently and repeatedly, you'll need to protect yourself. You can start by keeping good records. You should be doing that even if your narcissistic boss isn't a bully.

PROTECTING YOURSELF

Document Everything

Carefully document all exchanges and paperwork that prove you produced a product or idea. Record, verbatim, any insults or put-downs, with places, times, and the names of people present. If you've been instructed by a supervisor to do something

and it fails or turns out to be an error, keep a copy of all cor-
respondence. Do the same with threats for termination. In fact,
make a detailed note of any bullying behaviors. Make sure you
keep backups of supporting documents on an external hard
drive in case the copies at work are damaged, hacked, or stolen.
Also, keep your journal in a private notebook, or on your home
computer, not on the office computer; the company owns it—
and has access to whatever is on it.

If you feel you've been wronged—say, a colleague or boss
has taken credit for your valuable idea—don't rub their nose
in it. There's always the chance their theft wasn't intentional.
They simply might have forgotten the idea was yours. Let them
save face. Jane, for instance, re-sent her original email contain-
ing her idea with the following note:

"Yes, you may or may not remember, but this is what I suggested. See below. I'm glad you changed your mind about it and
we impressed the team with the idea! Please CC me when you
clarify the origin of the design."

Notice Jane used said "we" in her note. Language is everything. Even with documentation, try to include the first person
plural whenever possible. There's always room to prod people
toward the center of the spectrum even when you're protecting
yourself. With each of the remaining self-protective measures,
always try to remind the narcissist of your relationship.

Remain Focused on the Task

Instead of challenging the bad behavior directly, question its
relevance to successfully completing the task.

When Drew didn't shift his ways after her gentle correc-
tions, Jane's best approach was to keep him concentrated on
the task at hand. For example, when he started rattling off everything that she'd done wrong with a report, rather than de-

fending herself, she redirected him with the question: "Can you help me understand how this brings us closer to a solution? Did you have specific changes you had in mind that you can describe?"

The advantage of maintaining task focus is that it sets limits on bad behavior. This strategy has two elements: *querying* ("Can you help me understand how this helps us solve the problem?") and *requesting* ("What is it you're asking me to do?").

Be careful of your tone. It's easy to slide into a snide or snarky voice when you're feeling under attack. Try to remain calm and kind. Practice in front of a mirror if you have to.

Drew responded well to this approach. It reminded him he'd lost focus without taking him to task for the lapse.

Some examples:

- Your boss questions the quality of your analysis on a company's equity, and then moves on to questioning whether or not you have what it takes to be a success at work.

  Can you help me understand how this helps me with the problems you see? What is it you want changed?

- Your subordinate begins ranting about the abusive culture of the office (though you suspect he's largely to blame).

  Can you help me understand how this improves our work culture? What specific solutions did you have in mind?

- Your coworker starts blaming you for everything that's gone wrong on a project.

  Can you help me understand how this helps our project move forward? I'm not sure what it is you're asking for.

Block the Pass

If you're feeling helpless or overwhelmed after an interaction with anyone at work, you're likely at the end of an attempted hot-potato pass. You'll need to block it. In this approach, you encourage the narcissist to speak directly about the insecurities they're trying to get rid of, but again, in a collaborative tone.

Start by reflecting on your own feelings. That should give you a sense of what emotions are being fobbed off on you. Do you feel helpless? Ineffective? Under the gun? See if you can invite your boss, coworker, or subordinate to admit these feelings or some version of them by explicitly naming them on both your behalf.

Jane, for example, learned to speculate about Drew's own worries whenever he became hypercritical. Once, when he began questioning her commitment to their project because she couldn't stay late (she had a therapy appointment), she said: "I get it. We're really under scrutiny here, and you're especially feeling it today. What's got you more on edge right now? Maybe we can prioritize the most pressing tasks." Drew, to his credit, did manage to admit he felt on edge, which meant they could at least spend time talking about the pressure he'd been getting from upper management about certain tasks.

Block the pass is an important test. If the person can't admit their behavior reflects some underlying insecurity (without having to call it that), they might be too rigid to shift back to the center of the spectrum.

Other examples of responses to situations:

- When your boss implies your work is substandard (but offers no specific examples) and you're starting to feel like a failure:

You seem to be more anxious about the success of the project today. Have you heard something that has you worried?

- When your subordinate complains about work conditions after a poor evaluation and you're feeling attacked:

  I get the feeling you're feeling unfairly judged in the performance evaluation. Maybe we should talk about that directly?

- When your coworker questions your judgment on a decision you've made that everyone else loved, including your boss, and now you're questioning yourself:

  What's raising all these concerns for you right now? Has something shaken your confidence?

Of course, it's entirely possible that your boss or coworker is experiencing problems completely unrelated to work—say, in their marriage or within their family or in aspects of their life you're not privy to. But blocking the pass at least gives them a chance to reflect on how it's affecting their behavior. Maybe they'll even talk about what's stirring them up instead of putting you down.

NUDGING THE NARCISSIST

Catching Good Behavior

Most people picture confrontation as a showdown between the wrongdoer and the victim, more a form of punishment than conversation. And like most punishment, it tends to be disproportionate—a bit like the parent who screams, "Go to your room. You're on time-out!" It's more cathartic than pragmatic. Punishment brings a halt to behaviors, temporarily, but it doesn't encourage new ones.

In contrast, a more productive strategy is to look for moments when the person demonstrates *better* behavior and underscore them. Nudging narcissists to center means focusing on moments when they show some capacity for collaboration, interest in other people, or concern for the happiness of those around them—in short, whenever they behave more communally.

Drew frequently let his ambition and drive to be admired get in the way of his concern for others. But he also occasionally praised the quality of employees' work or asked about their home lives. Catching a high-spectrum narcissist being good comes down to highlighting these moments with appreciation. Jane, for example, took note when Drew, in a good mood, asked her opinion about a decision they were facing. Before answering, she tied his behavior to their success as a team: "I appreciate your asking me, Drew! Whenever you do that, I feel like a more valuable team member and that makes me want to work even harder to help us be a success."

Does this turn you into a sycophant? No. There's a difference between sucking up and giving legitimate praise. You're not making blanket endorsements of their conduct, only singling out specific instances of good behavior. Constant flattery is a no-no; it only feeds unhealthy narcissism. If you have nothing nice to say to a highly narcissistic boss, don't say anything at all. But if you see an action you like, make sure you point it out, clearly and sincerely (or you may never see it again).

To provide the proper nudge, draw attention to communal behavior and *always* tie it to workplace success:

To the boss:

Thanks for the great feedback! Hearing when I've done well inspires me to push even harder.

I always appreciate it when you ask how I'm doing. I get a little boost of energy to keep going.

To a subordinate:

I'm glad to hear you being so collaborative with the other team members; that's our best shot at having a big win.

It was kind of you to ask Jane if her mother was OK. Those simple acts of caring help everyone feel better and boost productivity.

To a coworker:

Thanks for the support in that meeting. When we're working together like this, we're bound to deliver a great idea.

Thanks for asking me if I'd like anything when you went out for coffee. Helps me push harder for the finish line when I feel like you've got my back.

Contrasting Good and Bad Behavior

Contrasting is much the same as catching, except that you're describing the past and the present at the same time. Noting bad behavior becomes far more effective when it's paired with some recollection of more communal behavior (assuming you've caught any). Jane, for example, when faced with one of Drew's suddenly "high priority items," simply described her experience:

I really appreciated it when we checked in at the start of the day last week (catching good behavior). I felt good about our team and brought my most creative thoughts to our project. I noticed we didn't do that today and it's definitely sapped my energy a little (contrasting bad behavior). I'd love to go back to how we did it last week.

Notice how often Jane uses the plural (communal) pronouns. Try to speak in collective language; use *we, our,* and *us.* Lapsing into *you* and *I* language isn't just less collaborative; it's likely to keep the person stuck in a narcissistic mind-set. If you end up using *you* at all, save it for behavior you're trying to reinforce. It's not a bad thing for high-spectrum narcissists to take ownership of their efforts to think about other people.

Some more examples:

To the boss:

I had such a great experience on our team last week when we left time for everyone to contribute. Today, we had less of a chance and I felt a lot less hopeful about the project. Can we try to do it the same way as last week?

I noticed on the last project, you discussed the errors calmly, and everyone, including me, had a much easier time problem solving than today, when it felt tense. How can we get back to the problem solving you helped us with last week?

To a subordinate:

Morale was a lot higher here last Friday, when your team discussion left space for people to talk about the options together, which was great. Today spirits seemed to flag. How can we work on maintaining your more open approach going forward?

Your enthusiasm brought out the best in everyone yesterday because you asked other people questions and showed an interest in their answers. How can we stay on track with that for the rest of the week?

Becoming Assertive

This is the most specific strategy of all the interventions. You're clearly stating what's wrong and what needs to change. Contrary to most recommendations, it's probably not the best place to start in the workplace. It involves the most emotional risk on your part, because you're saying more about what you feel. The closer your emotional connection is with your colleagues, the more appropriate this is. But most work relationships are formalized enough that deeper feeling statements can seem out of place.

Moreover, some narcissists might use your feelings against you, by shaming you or disclosing them to other people. This is abusive, of course, but it's not uncommon with more consciously manipulative and sadistic narcissists. You shouldn't share feelings until you have a clear sense of where someone falls in the spectrum. For all these reasons, I've placed being assertive last. Start higher up on the list and work your way down.

To be assertive, your statements should always include what I call the *ABCs*:

- *A is for affect, aka feeling.* Feeling statements use the word *I* liberally, as in *I'm feeling uncomfortable, uneasy, unhappy.* You can also use stronger words like *sad, afraid, scared,* but since you're usually not in a friendship or romantic relationship with the person you're speaking to, vaguer, less intense emotional language might be better. Follow your gut on that one. The main goal is to describe your experience only. Never use *you* in this step. Some examples: *I'm nervous; I shut down and can't think; I feel on edge.*

- *B is for behavior.* This is the experience, interaction, or action that causes the feelings. For example: *When you raise your voice; When I hear only criticism; When you sound sarcastic; When you cut me off midsentence.*

- *C is for correction.* This refers to the change you're seeking. Proper assertiveness always involves a request of some kind. It's a form of coaching. You're telling the listener what they need to do to improve interactions. Examples: *Can you lower your voice?; Can you tell me what steps you want taken?; Can you use a kinder tone?*

Bear in mind that tone in assertiveness statements is important. If you toss out your comments like rocks, the listener's bound to get defensive (and angry). Try to speak from a softer place, just as you did with empathy prompts in Chapter 8.

Jane used assertiveness with Drew once, after he continued to keep her late with questions and no other nudge seemed to reach him.

> *Drew, I feel uncomfortable when I say I have to leave and you keep asking questions. Can we agree that when I've told you I need to leave, we end the conversation?*

Other examples:
To the boss:

> *I feel unhappy the rest of the day when you criticize me in front of the entire group. Can you save your feedback for one-on-one meetings?*

To a subordinate:

I get nervous about how you're handling stress in the work-place when you flood me with e-mails about all the problems. From now on, I need you to prioritize which messages are important and which aren't.

Note: If you have to assert yourself with someone who works for you, consider taking administrative action, like a performance plan. It's a bad sign if your authority isn't enough to keep someone in line.

To a coworker:

I feel on edge when you argue with all my suggestions. Can we move to a more collaborative approach for the sake of the team?

EVALUATING THE RESULTS

It's helpful to set goals before you begin to employ these strategies. Having something to measure your success against will help you assess whether or not progress is being made. Your goals don't have to be grand or even specific—at least not right away. Simply ask yourself: What signs would indicate that the situation is improving for you? Is it being happier at work? Feeling more valued by coworkers? Having your opinion heard in meetings?

You needn't decide on all your goals before you start using the above strategies. Protecting yourself first might even help you clarify what's bothering you most. Jane, for example, only began to realize how much Drew consistently downplayed her contributions once she started documenting all her ideas. She recognized that if Drew's behavior didn't change, she couldn't possibly stay with the company. That knowledge helped her focus her efforts; she paid much closer attention to when Drew did—and didn't—give her credit. "I need him to recognize my work," she explained. "I'm not putting up with anything less."

Nevertheless, from the beginning, Jane's overall goal remained simple: to feel more comfortable going to work. She'd felt ill at ease at the office for months. Changing that was her top priority, regardless of whatever else improved. If you're like her, you may need to start with a simple, general goal as well.

What changes are you hoping for in your workplace? Make a list. Successful outcomes might include:

- You're no longer afraid to go to work
- You're sick less frequently
- You feel more creative
- You feel more valued
- You can stand up for yourself
- You feel emotionally safe (less likely to be unfairly criticized or insulted)

Alternatively, your measure might be more specific: having your work recognized, experiencing more reasonable job demands, or getting more consistent and fair compensation. Many people just want to be free from put-downs and yelling. Doubtless, if, like Jane, you've considered quitting, the situation's probably become dire and you're encountering outright bullying. You're probably looking for any sign of hope.

GOING HIGHER

If there isn't a sign of improvement, it might be time to appeal to a supervisor or human resource director.

But before you do, take a look at the file you've been keeping. It should include numerous examples of bad behavior by now. As you're reviewing evidence, speak to other people to get a reality check ("Does Drew ever make you feel the same way?"). If other employees describe a similar experience, you have a better chance of supervisors taking your concerns seriously. You might even ask other employees you trust—especially close friends—if they'd be willing to join you in your efforts.

Bear in mind, however, that this is an extremely risky move. You'll want to have a very clear sense of how supportive the company is about complaints and concerns before you go above your boss.

Unfortunately, going through official channels often doesn't work. Human resource departments often have the interests of management and the company foremost and either consciously or unconsciously find ways to dismiss the problem.

An informal workplace bullying study conducted in 2008 asked 400 respondents what their employer did when they reported problems. The results:

- 1.7% conducted a fair investigation and protected the target with punitive measures against the bully.
- 6.2% conducted a fair investigation with punitive measures for the bully but no protection for the target.
- 8.7% conducted an unfair investigation with no punitive measure for the bully.
- 31% conducted an inadequate/unfair investigation with no punitive measures for the bully, but plenty for the target.

- 12.8% did nothing or ignored the problem with no consequences for anyone, bully or target.
- 15.7% did nothing, but retaliated against the target for reporting. Target remained employed.
- 24% of employers did nothing, except fire the target.

Faced with these dismal statistics, psychologists Gary and Ruth Namie strongly caution, "Do not trust HR. They work for management and *are* management."

Some larger companies are taking problems seriously. They may offer consultations with an ombudsperson, someone who can hear complaints, advise, and provide counsel about how to proceed within a specific organization. Ombuds have intimate knowledge of the system and they're often the perfect person for groups of employees to see to document, assess, and help intervene with problems. Some even provide anonymous feedback to the executives of a company about the problems they're hearing.

Most ombuds offer confidentiality; just like therapists, they can't speak to anyone else without a written release. Nevertheless, always ask for a written statement from the ombud of all the limits and conditions for privacy, so you can review them and ask questions. Most workplaces provide them as handouts. Remember, though, that even therapy records are discoverable (opened up by courts) under certain circumstances. You can count on your ombud records being opened if you go the adversarial route by, say, filing a lawsuit. Consider that when you decide what to talk about. Don't share mental health history (such as depression or anxiety), for example, if you think you might pursue legal action unless you're comfortable with the information being discussed in court.

In considering appealing to the organization, you'll also want to assess the degree of cultural narcissism in a company. Management professor Andrew Dubrin, of the Rochester Institute of Technology, considers an organization narcissistic when it is "self-absorbed and suffers from delusions of grandiosity." Pay close attention to signs of profligate spending on meetings and celebrations, elaborate high-end office furniture, and CEOs with a cult-like following. Organizations that encourage and thrive on narcissism aren't likely to see entitlement and exploitative behavior as a problem; they might even frame it as "hunger for success."

The premier example is Enron, the one-time Wall Street darling that collapsed in 2001 due to rampant financial fraud. The company was known for its lavish parties, luxurious offices, and boastful executives. When it failed, employees were left penniless and executives, from the top down, didn't seem to give a fig about their plight. Your chances of being heard, let alone protected, are much lower in a place that rewards high-spectrum habits.

If you've taken your appeals higher and your supervisor or HR listened carefully to your concerns and even intervened on your behalf, you're in great shape. Keep working with them and providing updates on the situation.

If you've tried interventions at every level and the supervisor or system is unresponsive to your needs, you've truly done all you can, and you're in much the same position as the partner or friend of a narcissist who can't break their addiction. Your needs aren't likely to be met. The system, itself, may be stuck in an addictive cycle, which means the narcissist is merely a symptom of a larger problem.

Leaving a job can be as painful as leaving a relationship. And in troubled economic times, it can feel impossible. But if you've tried to make things better and still feel miserable, choosing to stay likely means continued misery. That's because your happiness isn't in your hands anymore, but the company or employer you work for. That's when it's time to take control—and leave.

PART IV

PROMOTING HEALTHY NARCISSISM

10

—

ADVICE FOR PARENTS

RAISING A CONFIDENT, CARING CHILD

Trish had been seeing me to help with stress from her job. About six months into therapy, our conversation shifted to her six-year-old son, Tommy. She'd just received a letter from his school that concerned her.

"He's been putting other kids down," Trish explained sheepishly. She grimaced, her eyes darting to the books on my shelf, uneasy with what she was about to share. "It started with little things. Correcting people's speech came first," she said, smiling. "Imagine a six-year-old commenting on grammar like some little schoolteacher. Half the time he was wrong, which somehow made it worse." She shook her head, her smile collapsing. She handed me the letter.

In the latest incident, Tommy had marched up to a boy on the playground, younger and quieter than most of the other kids, and announced his hat was ugly.

"The boy cried and ran inside," she explained quietly. "His father had given him that hat, and passed away a few weeks later. Tommy didn't know—how could he?—but that's beside the point."

"How long has this been happening?"

"A year or so, I'd say." She suddenly leaned back into her chair, her eyes narrowed in concentration, making a connection she hadn't realized. "I guess it was right after we enrolled him in a private school."

Tommy had shown signs of being gifted early on. By age three, he was reading books and speaking in paragraphs; by four, he'd often hunker down at the kitchen table, papers strewn around him, solving math problems for fun. Trish, a lawyer, and her husband, Brad, a physician, wanted to give him the best education possible, so they moved him from a public to a private school, hoping he'd receive more attention there. His new teachers applauded his outgoing nature, but some were concerned that he was also "impulsive and dramatic," blurting out silly answers or strutting about pretending to be "the boss"—a character he invented who'd cured cancer and now traveled the world solving other problems.

"His grandfather died of colon cancer, and I think 'the boss' was his way of dealing with it," Trish explained. "But he can get quite pushy, and I'm worried his grandmother, Margaret, has been encouraging all this. She spends a lot of time bragging about his school achievements, telling him how brilliant he is. She never puts him in time-out or corrects him for anything, even when he hits his sister, Jill. My whole life, my mother seemed convinced she could do no wrong, and now she acts like her grandson's perfect, too.

"What happens when you talk to her about it?" I asked.

"There is no talking to her. I tell her to praise him for other things, too, like how sweet he can be with Jill or how hard he works. But she ignores me—she always has. The irony is she did nothing but criticize me and my sister growing up." She clenched her jaw, grinding her teeth while she thought.

"I'm determined not to become as critical as my own mother was, but sometimes, when I'm really angry, I don't know how to handle Tommy. The other day, he insulted his sister for about the fourth time that afternoon—he said her drawing looked silly—and I just lost it. I told him he couldn't have a playdate until he apologized, then I said no one's going to want to be friends with him if he acts this way." She tapped her feet and sighed, her anxiety building. "He loves attention and hamming it up and he's never afraid to say what's on his mind. I don't want him to lose that, but how do I make sure he doesn't become as big a narcissist as my mother?"

Trish is right to be worried. As you learned in Chapter 5, some children show early signs of unhealthy narcissism—and Tommy displayed quite a few. For one, his classmates had grown tired of the way he bragged about finishing his work first and wagged his finger at them when they didn't know an answer. His arrogance left Trish in a quandary, one that many parents face. She couldn't ignore his misbehavior (his grandmother's approach). But she also didn't want to get into constant, angry battles as he violated one rule after another. She wanted Tommy to feel nurtured and loved, while nudging him in the *right* direction— toward the center of the spectrum, where creativity, empathy, ambition, and self-assurance flourish without all the arrogance and put-downs. And she didn't want to harm him in the process.

Trish's dilemma is one shared by many parents. Early experiences, we all know, can set children on the right—or wrong—

path for the rest of their lives. Trish and parents like her live in terror that they're raising obnoxious narcissists. But what they don't realize is that they also have the power to promote *healthy* narcissism. So what's the magic approach, the "perfect" parenting style that brings out the best kind of narcissism and discourages the worst? To get a better sense of this, let's first take a close look at the four basic approaches to parenting.

PARENTING STYLES

Parenting has two major components: warmth and control. Warmth is the caring, love, and nurturance we show our children; control is the direction, monitoring, and guidance we provide. Children need both of these components, but the right balance is crucial; too much of one—or one without the other—prevents them from thriving. In fact, it's the balance of these two factors that creates the four parenting styles, each of which has a different effect on a child's degree of narcissism:

Authoritarian

Authoritarian parents are highly controlling and often feel cold and emotionally distant to their children. Authoritarianism is heavy on demands, without being particularly flexible or responsive to a child's needs. It also easily becomes abuse; it's a short step from control to cruelty.

Authoritarian parenting, nevertheless, can take a couple of forms that aren't overtly abusive. At the icy end is "tiger parenting," popularized in legal scholar Amy Chua's *Battle Hymn of the Tiger Mom*. According to Chua, tiger parents control their children's every waking moment to ensure their success. They *order* them to get A's and prevent them from seeing friends so they can work hard enough to earn top marks. There's no coddling and certainly no concern for a child's self-esteem. Researchers who've studied tiger parenting, which is thankfully rare, find most youngsters reared in this atmosphere are unhappy, anxious, depressed, socially inept, and ironically, underachievers.

At the warmer end of authoritarianism there's "helicopter parenting," a term that's used liberally—and for the most part incorrectly. Some people think that helicopter parenting is defined by extreme involvement in their kids' lives—for example, having daily contact with college-age children, helping them pick their course majors or develop term paper topics—all of which, according to research, might be associated with a host of benefits, including happiness and better grades. Psychologists, however, define helicopter parenting more precisely, reserving the term for a pattern of excessive control and *interference*. College students who have been reared this way agree with statements like "My mother monitors my exercise schedule" and "If I'm having an issue with my roommate, my mother would try to intervene." Helicopter parents aren't frigid but their constant interference makes them seem coldly indifferent to their child's feelings. The results are much the same as with tiger parenting: children who are miserable, anxious, and depressed. Luckily, this form of authoritarian parenting is rare, too.

Growing up in an authoritarian household is a bit like living in a police state; children are constantly shaped and controlled, forced to submit to the will of the all-seeing, unquestioned parental authority. Children raised this way often become hobbled in their ability to comfortably depend on anyone, leaving them at the greatest risk of becoming echoist or, especially, narcissistic adults. This style turns a person into a performance.

Authoritarian Parents

Believe children should be seen and not heard.

Don't allow their children to express anger toward them.

Have extremely strict, well-established (though not necessarily consistent) rules.

Don't allow their children to question them.

Usually express very little warmth.

Permissive *or* Indulgent

Parents with this style tend to be warm and undercontrolling.
When Trish felt her most exasperated—as she did after receiving the note from her son's school—she slipped into permissive
parenting. In part, this was a reaction to her own upbringing.
She lived in fear of becoming like her authoritarian mother,
Margaret, who called her children ugly names, lashed them
with a belt when they broke her rules, and even threatened to
"give them away" when she got angry enough. Trish had spent
years in therapy, learning to break her method of coping with
her mother, that is, by echoing her narcissism. Trish still had a
tendency to shut down whenever she became angry and would
withdraw when Tommy became cruel.

A few weeks earlier, when her husband was away on a business trip, Trish and her children spent a long day together at
the zoo. Tommy had become exhausted, but when they got
home he insisted on marching around barking orders at his
sister. "You need to do your homework!" he bellowed at Jill,
then raced into another room and started rifling through her
backpack (presumably to find her assignments). "The boss"
had come out in full force. Trish felt as fatigued as everyone,
and after half an hour of attempting to redirect Tommy's focus,
she gave up. "I'm going to bed," she said, and closed her door.
Tommy continued making a commotion until he grew too tired

to carry on. Trish emerged from her room, then, to carry her son to his bed and sweetly sing him to sleep.

We can all sympathize with Trish's weariness and sometimes giving ourselves a time-out (one way of framing her choice) is the smartest move we can make. But, typically, Trish never came back to the problem the next day. "I woke up fuming," she explained. "And I just wanted to have a nice morning."

Permissive parenting, in a nutshell, is all warmth and no direction. At Tommy's age, he needs more guidance from his parents, not less. Without it, he might come to believe that he's so special he doesn't need *any* rules at all. We should expect more self-control as children get older, but permissive parents tend to stick with their warm, directionless approach even when their child has clearly stepped out of line. That fosters unhealthy narcissism, too, especially if parents remain permissive after a kid like Tommy reaches adolescence.

Permissive Parents

Feel children should have time to think and daydream or even loaf.

Let children make many decisions for themselves.

Provide comfort, but few rules.

Find it difficult to punish their children.

Often let bad behavior go or rationalize it ("boys will be boys").

Indifferent *or* Neglectful

This parenting style is both cold and undercontrolling. Trish's new neighbor, Monica, had a twelve-year-old son named Eric who'd quickly gained a reputation as the neighborhood bully. He brandished sticks and cursed at younger children, overturned trash bins, and played loud, pounding music at all hours.

Monica, a divorced single mom, worked long hours and relied on babysitters who failed to monitor Eric, much less control him. When Monica took over in the evenings and on weekends, things weren't any better. When anyone confronted her, she'd nod and smile and say, "Yes, of course he needs discipline," but as far as anyone could tell, she never provided it. The few interactions between the pair that Trish had witnessed seemed distant and strained. "She'll just stand there, tapping at her phone, even when Eric's tugging at her sleeves."

Monica is a textbook example of indifferent parenting and, given her complete emotional absence, it's not surprising that Eric seemed well on his way to juvenile delinquency and extreme narcissism.

Indifferent Parents
Expect their children to handle problems by themselves.
Sometimes forget their promises to their children.
Push their children to be independent.
Often don't know where their children are.
Express little love and affection.

Authoritative

Authoritative parents blend warmth with discipline. They gently guide, with love and affection, but adjust their expectations and rules to the age and needs of the individual child. Part of effective parenting is knowing when to step in and when to step back. When children are infants, they can't do anything for themselves, so we do everything for them. As they grow into toddlers, we still take care of their needs but can move back a bit and loosen the reins. A three-year-old may not be able to tie her shoelaces, but she can pull on her pants. As they grow older, we allow them more freedom. A 12-year-old coming home from school to an empty house may have to call a parent at work to check in—it's comforting to both child and parent; but the same mandatory voice or text updates aren't attuned to the needs of a healthy 16-year-old, who's in the process of developing an independent sense of identity.

Reins continue to slacken as a child matures, but they can—and should—tighten back up when a parent senses the child is in danger. A teenage boy who starts staying out to the wee hours of the morning, partying hard, drinking alcohol, and smoking pot, requires urgent and direct guidance—and a curfew. And he'll likely require a lot more guidance than his younger sister who sticks to her own curfew of 10:00 p.m. without being reminded.

Combining warmth with the right degree of control, at the right time, makes children feel safe and secure and creates happy, successful adults with a healthy degree of narcissism.

Authoritative Parents

Respect their children's opinions and encourage them to express their feelings.

Talk things over and reason with children when they misbehave (depending on age).

Include their children's preferences in future plans.

Trust their children to behave properly when they're older, even when not with them.

Adjust their demands to meet the age and emotional maturity of the child.

Listen closely to their child's needs and feelings and try to understand them.

Trish learned, through our work and the help of a child therapist, to find the constructive balance between affection and monitoring to help Tommy live closer to the center of the narcissism spectrum. She started by developing the tools of authoritative parenting.

BECOMING AN AUTHORITATIVE PARENT

Authoritative parenting strategies all share one thing in common: they teach children to consider their impact on the people around them. The reason permissive parenting, for all its warmth, fails to guarantee healthy narcissism, is that it doesn't ask kids to consider other people at all. Authoritarian parenting leaves them so hemmed in they barely feel like a person. Healthy narcissism is all about knowing your own voice while still hearing others. And that's exactly what authoritative parenting teaches children to do. With that in mind, I've put together a list of strategies, based on decades of research, to help you promote healthy narcissism in your kids.

Practice Firm Empathy

Many parents confuse close listening and empathy with agreement. But sometimes parents need to be firm and stick to their rules despite their child's feelings. This gets tricky when a child feels sad or angry or scared.

Tommy, for example, became afraid of traveling by car after one of his grandparents passed away. "We took a long road trip to Orlando, Florida, to see his grandfather," Trish explained. "But he died shortly after we arrived, and now Tommy's convinced that long drives are dangerous." In the past, Trish and her family often took day trips to the country on the weekend. Now they barely went at all. "Tommy throws a tantrum anytime we have to go somewhere more than an hour away."

Trish and her husband had spent hours—Trish had been especially patient—speaking to their son about his fears. "I asked him, 'Are you afraid someone might die if we take a long trip?' Tommy nodded yes, but we couldn't get him to say much more about it. He just lay in bed, curled up, sobbing. Why push if he's that sad and scared? My mother used to push me all the time. I don't want to become like her."

The problem with this point of view is that it assumes that demanding something of children when they're upset is somehow thoughtless or selfish. That's certainly true if a parent's response to the child's fear is along the lines of "We're going whether you want to or not!" But it's not the case if you remain empathic while sticking with your plans.

Trish learned to approach Tommy with firm empathy, combining her well-developed talent for listening with a refusal to change their plans. "I know it's scary for you, Tommy," she told her son. "It's awful to be so afraid. But I know fear only gets worse if you stop doing what you're afraid of. What do you think will help you feel safer taking the trip? Do you want to bring your giraffe, Pokey?"

Firm empathy is deeply caring. It's important to recognize— and hear—when your child's afraid. But to work around their fears, to avoid further upset, guarantees they'll live a life of fear. While it's tempting to do this sometimes, we have to recognize that when we do, we're not really taking care of our children but ourselves. And that's another path to narcissistic addiction.

As parents, it's our job not just to understand our children but also to help them grow. Tommy learned he could depend on Trish to help him feel better both in a single moment, and— as he grew more comfortable with longer trips—across time. Tommy learned that his needs and feelings, while important, weren't so special that they trumped everyone else's. That's what kept him closer to the center of the spectrum.

Catch the Good

We live with our children, day in and day out, witnessing countless interactions, often in the span of a few hours. That gives us plenty of opportunity to apprehend their best behaviors if we're watching closely. When it comes to healthy narcissism, that means catching your child offering help, expressing vulnerable feelings, asking about other people, or apologizing.

Highlighting and rewarding caring and consideration is the best way to encourage new desired behaviors. It's neither necessary nor effective to rush in and criticize every instance of selfishness or grandiosity. As we've seen, research suggests that strengthening any communal thought, feeling, or behavior can help shift narcissists back toward the center of the spectrum. That means that commenting on the most touching moments— say, when your child kisses a tearful sibling's forehead or holds the hand of a frightened friend—does more than reward love and kindness. It reduces the sense of entitlement, helping foster healthy narcissism.

It helps to have a list of behaviors to track. Trish kept one for Tommy that included:

- Expresses appreciation
- Asks for help instead of getting angry
- Apologizes
- Says something supportive to his sister ("I hope you feel better")
- Names a softer feeling like sadness or fear

Once, for example, after Tommy trashed his sister's painting ("Why do you always do those dumb flowers?"), he caught himself within moments. "I'm sorry," he said. "That was a put-down." He walked up to Jill and inspected her picture.

"That was great, Tommy," Trish exclaimed. "You realized you used a put-down and said you were sorry! That's so important. Everyone feels better when you're willing to say you made a mistake. What else can you say to help her feel better now?"

"I could say something nice." He paused to reflect. "I like the colors and I'm glad you paint. The pictures make me feel happy."

Trish also took plenty of opportunities to *contrast* Tommy's behaviors. If you recall, in contrasting, you mention an instance of better behavior instead of dwelling on a moment of insensitivity. The day after Tommy apologized for his put-down, he slipped into the old pattern that Trish called "hit-and-run." He'd come home angry after a teacher put him in time-out, and handled his bad mood by telling his sister her collage "looked messy." Then he walked away. Trish caught up to him and used contrasting.

"That was so great yesterday when you apologized to your sister. She felt better and the two of you had a good time. Do you remember how you did that? Maybe it would help you feel better, too."

Model Vulnerability

Remember that the antidote to narcissistic addiction is the ability to openly acknowledge our fears, sadness, loneliness, and other softer feelings—and trust people to hear them when we share. Teaching children to do this is important in building their capacity for healthy intimacy. And there's no better way to teach than by example.

Trish lived in terror of lashing out at Tommy, and that fear made her pull back at times when his behavior veered out of control. She learned, instead, to state her sadness and fears. "Tommy, I feel so sad right now about how you're treating your sister and I'm so worried about you. I can't figure out how to help you with this right now, but I plan to think about it and come up with a consequence for your being so hurtful. But right now I'm just too upset to do that."

Tommy's father, Ian, also learned to say, "I'm feeling tense all over right now because of how you're acting and I'm afraid I'll get really angry if you keep it up. You need to take a time out."

These statements, as simple as they sound, speak volumes to your child. They send the message that you care, but that your feelings matter, too.

Set Limits

Certain behaviors, like hitting and hair pulling, should be off limits no matter what the child's age. But emotional cruelty—put downs, insults, name-calling—should also have consequences. For brutality and callousness, most experts recommend a time-out, where you remove the child from the current situation and set a timer for a specified period of time (a minute for each year of age is a good rule of thumb).

One of my favorite books for managing problem behavior with time-outs and other consequences is psychologist Tom Phelan's *1-2-3 Magic*. I like it because it's simple and easy to use even when you feel like you're going to explode. At the first instance of problem behavior, you simply count 1, calmly and slowly without raising your voice. At the next incident, you count 2, in the same manner. It doesn't even have to be the same behavior to receive a count—it could be any action you want your child to stop doing. If a child reaches 3 within a half hour window, he or she gets a time-out (or another consequence, depending on the behavior). With extreme behavior, like hitting, you can go straight to 3. Just make sure you explain how the system works before you start using it—and be consistent once you do. In other words, be predictable.

Predictability is a crucial ingredient in loving relationships. Children need to know *why* you expect certain behaviors and punish others. You can't simply impose rules and limits without explaining what they're for, even if the explanation is as simple as "I need to make sure everyone in the house is safe and this rule makes that possible." When children know what's going to happen, and why, they feel safer because their environment makes sense to them—and so do you.

As children become older, limits might take different forms, like a set consequence for broken rules. For an adolescent, the rule might be to come home by 11:00 p.m. The consequence for not meeting curfew might be no car for a week. Use any form of limit setting you like (there are countless books on it), but whichever approach you use, you'll do best to think of limits like walls. They block children from continuing in the wrong direction, but they also protect them.

When I was chief psychologist on a psychiatric unit, we became familiar with a phenomenon where agitated patients would escalate their threats ("I'm going to throw this chair!"), even after being informed that the outbursts would land them in restraints and "the quiet room." Within minutes of being restrained, they'd often calm down, as if suddenly relieved. I shared that impression with one of our patients, a tall, boyish-looking man who'd been hospitalized for months. "I feel safer there," he explained. "The walls hug me."

Setting limits with our kids works the same way. When they see their rage won't be allowed to hurt others or themselves, their world becomes safer. They feel hugged. Limits are a form of love.

Coach Your Child

Tell your child what to do, instead of what not to do. Many parents rely on laying down limits or prohibitions, but neglect to teach their children better behaviors. Children generally don't misbehave out of the blue. Very often, their actions stem from not knowing what to do with their feelings. Aggression comes easily to us as humans. It's an ancient wired-in response, a throwback to the time before language. Perhaps that's why the most aggressive children are often the ones who have trouble communicating in words. It's our job as parents to teach them how to express their feelings in positive ways. Instead of only administering consequences for bad conduct, take the time to explain better ways of handling uncomfortable feelings and situations.

When Tommy marched around playing "the boss" after a distressing day at school, Trish would take him aside and gently redirect him. "Tommy, I know you had a hard day at school and you had to miss recess. It helps to say, 'I'm sad or angry about what happened today.' Instead of bossing your sister around, can you tell me about your feelings?"

To develop empathy, young children have to be coached to name their feelings. That isn't something that comes naturally. They need to hear the language of emotion many times for it to sink in and make sense. One easy way to teach them is to name your own emotions when you feel them—and link them to your behavior. Example: "I'm feeling sad because Pepper, Aunt Helen's cat, died. That's why I'm quiet."

Another way is to help youngsters talk about their feelings. This can happen when they're upset. Try offering a few emotions and asking which one best fits what they're feeling: *Are you feeling sad, scared, or angry?* Alternatively, you can choose a more fun time, say, when reading favorite books, and explore the feelings and motives of beloved characters. Tommy loved a tale in which a boy sometimes got angry with his best friend. Trish asked, "What do you think he's feeling in his body? Have you felt that before? What do you think would help him feel better?"

In coaching, you'll want to make sure that any demands you make are appropriate to your child's age and maturity. Trish found it extremely helpful to talk with Tommy's therapist about how much emotional understanding should be expected of a six-year-old. Once she knew Tommy was lagging behind his peers, she stepped up her efforts at coaching him in managing his feelings.

Be Warm, but Respectful

While you're busy with the work of parenting, don't forget to stop and just enjoy your children. Hold them. Kiss them. Cuddle them. Curl up in a chair and read them a book at least once a week. This is all part of warm parenting. As they grow older, keep offering them lots of physical affection, but only if they're comfortable with it. Make it an invitation, not an expectation.

When my twin girls were around three, they often greeted me excitedly at the front door. "Daddy, Daddy," they'd squeal, as they came racing to the entrance to give me a hug and kiss. I looked forward to that greeting all day. One evening, as I unlocked the door, I heard the same mad scramble I'd grown used to. But when I entered, something unexpected happened. One daughter, Anya, simply stood there, blinking, a faint smile flickering across her face.

"No hug?" I asked

"No, no," she said, shaking her head. Her sister, Devin, jumped up and down behind her, eager to greet me.

"How about a kiss?" I said, turning one cheek toward her.

"No, no," Anya said again, shaking her head. At this point, I was going on pure instinct. The moment felt important to me, but I couldn't think why.

"How about a high five!" I said, beaming. Anya looked around, as if considering, and after a moment, squealed with delight.

"Yeah!" she squeaked, and leapt up to slap my hand. Devin raced up to give me her usual hug and kiss.

Only later did I realize that Anya had been exploring new ways of making contact with me, ones that were on her own terms, and testing my reaction. Would I insist on my hug, becoming angry or sad when she refused? Or would I adapt to her new way of connecting? Anya wasn't aware of any of this. What she'd been doing—experimenting with independence—was unconscious. But for me, the moment became a powerful metaphor.

As children grow, they need us to be there for them but they also need room to be who they are. They need to experiment with the space between themselves and the people they love. Sometimes they'll want hugs, which demand we stand close and embrace them. At other times they might want a little more space—a high five from a few feet away. If we rigidly insist on their loving us in just one way, they'll dutifully hand out hugs or kisses or high fives or whatever we seem to demand. But they'll also come to fear that they're special, in our eyes, only if they conform to what we want. In short, they'll learn to be special *for* us. There's no surer way to foster narcissistic addiction. Children literally bend themselves out of shape if they think it'll make us happy. They need love that much.

If your children want physical distance, give it to them (as long as they're safe), but tell them you'll be there when they want you closer. If they're not open to hugs, let them greet you on their own terms. If they don't feel like talking, tell them you're here when they do. If they want to retreat to their rooms, invite them to come out when they're ready. If your teens need space, insist on respect but don't force them to open up. Just be there.

Model Repair: Use "Re-Dos"

I often tell couples in therapy, "You can't get close enough to touch someone without stepping on their toes." We inevitably hurt the people we love. The key to happy relationships with our children—or with anyone, for that matter—isn't being perfect. It's having the courage to acknowledge when we screw up. That's repair work, and it's central to developing healthy narcissism.

Repair work means you always have a "re-do." What's a re-do? Ian was working in the garden one day, trying to fix the sprinkler system, when Tommy came home from school. Frustrated and immersed in his repairs, Ian grunted, "Hi, Tommy," and waved the boy off.

"Afterward, I felt terrible," Ian told me. "I realized this is exactly the wrong behavior to be modeling. So I came inside, and as soon as I saw him, I apologized."

"How?" I asked

"I told him, 'I'm sorry, Tommy. I didn't greet you with a happy hello because I was so stuck in my project. That probably didn't feel good. So let me try again. "Hi, Tommy! Welcome home," ' I said, and gave him a big hug."

That's a re-do. You acknowledge the mistake and try again. If you teach your kids how to do that, they'll learn that mistakes are a part of closeness. Repair and love go hand in hand.

Volunteer

Volunteering can promote healthy narcissism by teaching youngsters other ways to feel good about themselves besides achieving high grades or winning sports trophies. People who help the homeless or take care of sick animals or contribute to any group beyond themselves and their immediate family feel happier. Giving provides everyone with a rush.

If you want your children to feel good about themselves in healthy ways, build some charity work into your time together. Have younger kids go through their toy box or closet and pick out games and dolls or clothing they're ready to donate. Take older children to a soup kitchen or a homeless shelter to help serve food or read a story to toddlers. Talk to them about the experience and ask them what they think it would feel like to be homeless. Encourage them to learn the stories of the people they meet. Expose them to the world beyond the one they normally see everyday.

One of the most powerful experiences Trish and Tommy and her family had together involved bringing Christmas cookies to kids in a shelter.

"Tommy seemed a little confused at first," Trish said. "He asked me if they really lived there. I said, 'Yes, sweetie.'"

Tommy settled in a little more after he met six-year-old Mandy, who'd been at the shelter for several weeks with her mother. The two children sat in a corner and Tommy peppered Mandy with questions. "That's when Mandy started crying," Trish continued. "She told Tommy her father had been 'mean'—that's the way she described his abuse—and they had to leave home. Mandy was sad because she missed her house and her friends." Trish dabbed her eyes and sniffled. "Later in the car, Tommy said it's bad to be mean because it takes away from other people."

That's the power of helping other people. Even for children, it forces them to see the world from someone else's point of view. And they learn just how important connection and caring is not just in their family, but in the world. That's about the best way there is to guarantee a life in the center of the spectrum.

Parents aren't the only influence on children, of course. Thanks to the rise of digital media, we live in a brave new world full of opportunity to take center stage on a scale unprecedented in history. Thus far, we've heard dire predictions that social media, tellingly abbreviated SoMe, is a harbinger of a coming narcissistic apocalypse. It begs the question: Is there a way to promote healthy narcissism in the digital universe?

11

SoWe

THE HEALTHY USE OF SOCIAL MEDIA

"Oh, you have to be on Facebook and Twitter," my friend Ben bellowed over his cup of coffee. He gets loud when he talks about social media. "It's the best way to build an audience!" Ben, a SoMe veteran, had been using Twitter and Facebook for years. He'd offered to help me set up accounts and explain how everything worked.

"When you write an article," he went on, "you post it to your Facebook fan page and tweet it to followers so they can share it." He took a sip of his coffee, tapping at his laptop with one finger. "See, I've already been re-tweeted four times!" I stared blankly at the screen. "That's when people send your tweet to all *their* followers," he added, catching my mystified look.

Later, I hunkered down at my computer and set up my fan page, an act that generated two online identities: "Craig Malkin," the guy proudly posting about his twin daughters' latest neologism ("alackadactic," which, they told me, meant sunny), and "Dr. Craig Malkin," author, who'd just shared an article on overcoming jealousy with his "fans" (the readers and friends who'd elected to see all my latest blogs and information). That gave me two ways to feel inadequate. I could count how few comments my personal updates had accrued compared to everyone else's, then browse over to my professional page and see how little attention I was getting there, too.

I watched my Facebook feed with a mixture of fascination and horror. Most of the posts didn't appear to be news at all. They were more like snippets of daily life. "Kicking back on the couch, watching my new flat screen," wrote one friend. That drew fifteen likes. "Kids had their first snowball fight this morning," wrote another. That message got ten likes—and a comment. "Giants Suck!" typed a disgruntled football fan. That update garnered a whopping 86 likes and 40-plus comments.

At this point, I couldn't figure out the appeal of Facebook. I only knew that with all the likes and comments, Giants Suck! was a far more popular topic among my friends or fans than any of my posts, including my latest article. That rankled me for reasons I couldn't explain. SoMe was a strange new world, and all I knew is that I wanted it to recognize me in some way. And it hadn't.

My Twitter feed seemed no better. Those posts, restricted to 140 characters, occasionally carried real information (an actor or author's next public event) or served a worthy purpose (drumming up support for a cause) but most appeared to be blasts of banality ("My dog just threw up again #gross") or—even worse—a series of disguised sales pitches. Celebrity tweets about their "favorite" products are usually well-remunerated advertisements. Sadly, people often mistake these chirps for real conversation-starters, and it isn't unusual for them to draw hundreds of replies. And when the celebrities don't chirp back, it often leads to another imploring round of tweets. "@justintimberlake—please follow me back!" or "@taylorswift please reply!!"

Indeed, wherever I looked in social media, everyone seemed to be scrambling to be noticed. It didn't matter which side of social media we inhabited—whether we were the poster or the follower, we all longed to be recognized. For our talent. For our looks. For our wisdom. For a paycheck. For anything.

I quickly realized that social media is a *stage* for people seeking attention of one kind or another. And that's what gives it such power. It's why it pulls us in. With it, we can step into a virtual spotlight and share (or, create) our "story," no matter how big or how small. We can reach out to celebrities and hope that they "see" us. We can even develop a base of "fans" who love hearing what we have to say. For the seconds or minutes or days our updates remain visible, we feel important. Somebody, somewhere, is thinking about us. We feel *special*. Social media feeds our narcissism.

Over the next month, I grew more and more obsessed with my status in the world of social media. I felt warm all over when my "like" count rose. When it stayed flat, I felt uneasy, as if the universe had found me lacking. I grumbled in disappointment and flushed with envy when other people's posts drew more attention. I thought far too much about what to post and when, often spending hours crafting a single message.

Earlier that spring, I could have cared less about Facebook or Twitter. Now, suddenly, I couldn't stop checking my numbers. Feeling special in social media had quickly become an addiction. I knew I had to find my way back to the center of the narcissistic spectrum and I spent the better part of a year trying to do it.

Where did I begin? By trying to answer a question begging to be asked: Can the narcissism of social media *ever* be healthy?

IS SoMe SO BAD?

When I began searching for an answer, I soon found myself floating adrift in a sea of contradictions. No sooner did a study claim social media destroys our self-esteem than another appeared saying it boosts it. Some research concluded it merely expands our social lives, making it easier to connect with people we love; other studies claimed social media drives lonely people further into isolation or turns users into raging narcissists.

So what do we know for sure?

Well for one thing, it's a mistake to assume that all forms and uses of social media have the same impact.

Different platforms encourage different behaviors. Some make it easier to focus on appearance or fame. Some encourage conversation. Still others revolve around shared interests—say, your favorite music, through Spotify, or cherished images and articles, with Pinterest. As simple as these sharing sites are, they still provide a chance to bond with friends, as I soon discovered.

One day while browsing for new music on Spotify, I noticed, scrolling by on the right side of my screen, the tracks a friend of mine had been listening to. Because I appreciate his taste in so many things, I started listening, and I was hooked after the first song—a bright blend of bluegrass and jazz. My friend was delighted, of course. We messaged each other about it and later we exchanged rapid-fire text messages about our favorite tracks. We both felt good: our taste—and our friendship—had been affirmed.

It's hard to see the downside of putting ourselves on display when it leads to moments like the one my friend and I shared. But seeing someone's playlist is a far cry from interacting in profile-based platforms, like Facebook or Google+, where you can view whole chunks of friends' lives, spanning days or even years, and comment on what you see. And it's vastly different from microblogging sites, like Twitter or Tumblr, where if you speak to people at all, the conversation generally happens in short bursts (microbloggers sometimes share images and articles for months without ever exchanging a single word with the people who follow them).

All social media platforms have their own customs and rules. They look vastly different from one another. And they *feel* different when you use them. That's why instead of regarding them as tools, we should probably think of them more like countries or cultures. Asking if Facebook or Twitter causes narcissism or unhappiness is like asking if living in Russia or Iceland causes loneliness or cancer. It depends on where, in each country, you spend your time, but also, obviously, on what you do when you're there. Once I realized that, it became a lot easier to figure out how and why social media drives any of us up—or down—the narcissism spectrum.

SO FABULOUS: LOOK AT ME! LOOK AT YOU!

Based on what we know about human behavior offline, anything that takes us further away from authentic relationships is more likely to feed narcissistic addiction. That holds true in the digital world as well. It's all too easy to hide our vulnerabilities and trade empty show for true sharing—and that pushes people toward both ends of the narcissism spectrum.

Psychologists Brittany Gentile, of the University of Georgia, and Jean Twenge, of San Diego State University, designed an experiment focused on Myspace, a site where people often boast about their looks or social standing and post provocative photos. The researchers' goal: to find out if the primping and posing actually increases narcissism.

The team randomly assigned 20 male and 58 female students to spend 15 fifteen minutes either editing their Myspace profiles (presumably sprucing them up with flashy new snapshots or descriptions of themselves) or plotting routes from one campus building to another using Google maps. Then, they measured the students' narcissism. The result? The Myspace group yielded significantly higher scores. Fascinatingly, the Myspace profilers were also more likely than the Google mappers to agree with highly grandiose proclamations like "Everybody likes to hear my stories" and "I always know what I'm doing."

These findings have implications for all social media sites. We present our best selves in cyberspace. No matter who we are in real life, our social media avatar tends to be a cleaned-up version of the truth. We select our most flattering photos and cherry-pick information, sharing the things that show us at our brightest and happiest. When people "like" what they see, the boost to our self-esteem can be potent. But what the Myspace study suggests is that if we spend too much time prettying ourselves up, we can easily slip into vanity and self-obsession.

The obverse happens, too. When we spend lots of time examining other people's profiles and posts, we can damage our own healthy narcissism. In an ongoing study, professors of journalism Petya Eckler, of the University of Strathclyde, and Yusuf Kalyango, of Ohio University, surveyed 881 female college students, who spent an average of 80 minutes per day (many spent more) checking out their girlfriends' Facebook pages. The researchers found that the more time the women spent inspecting their friends' photos, the worse they felt about their own bodies, particularly if they wanted to lose weight.

The results make sense. Earlier studies have shown that the more time women spend leafing through fashion and beauty magazines, the worse they feel about their bodies. But comparing yourself to retouched photos of models and actors is one thing. Comparing yourself to friends is even more damaging. Our most intense feelings of envy aren't stirred by distant celebrities but by the people we know. Today, observes Eckler, more and more young women Photoshop their pictures before posting, using apps like SkinneePix, which shamelessly boasts it can help you "edit your Selfies to look 5, 10 or 15 lbs. skinnier in two quick clicks" on your smartphone. The bodies the women in the study were comparing themselves to may have been no more real than the ones gracing the pages of the average fashion magazine.

Facebook is a giant compendium of your friends looking happy or hot, and it hits women—and men—hard. Everyone's life looks better, shinier, more full of enviable debauchery, lavish vacations, thoughtful romantic partners, and perfect, smiling families. Sociologist Hui-Tzu Grace Chou and Nicholas Edge, BS, of Utah Valley University, asked 425 male and female undergraduates how many years they'd been using Facebook and how many hours a week they typically spent on the site. They also asked them to rate, on a scale of 1 to 10, the extent to which they agreed with various statements, including "Many of my friends have a better life than me," "Many of my friends are happier than me," and "Life is fair." Once again, the more hours the students logged on Facebook (and the longer they'd been using it), the worse they felt about themselves. Not only were they more likely to agree that their friends were happier and had better lives, they also tended to think life was unfair.

This study didn't monitor specific activity, but it's a good bet that more time meant more comparisons. That assumption found strong support in something else the authors discovered: people felt even worse when they had more "friends they didn't know personally." Why? Because they had no chance to correct the rosy impressions they saw on Facebook with *real* information about the lives of their "friends."

Social media has granted us unprecedented control over our personal narratives. With every click—with each image or comment we post—we shape the story of our lives. We need to be careful about what story we tell. And we need to be mindful of how we view other peoples' stories. The lesson of research regarding social media is that we inevitably suffer when we base our personal happiness on what we see there.

But there is a second lesson we can draw: We can use social media to improve our own and each other's lives.

SOWE: THE HEALTHY USE OF SoMe

After they completed the Myspace study, Gentile and her colleagues ran a second experiment, this time with Facebook: they randomly assigned a group of students to spend 15 fifteen minutes either editing their page or fiddling with Google maps. Like the subjects in the Myspace study, the students on Facebook felt better about themselves, but here's the difference: their self-esteem got a lift; their narcissism didn't. Why? The authors concluded that Facebook fosters a *community* experience—people reaching out and supporting one another—whereas Myspace nurtures individual exhibitionism. The study shows that using social media the right way, that is, by emphasizing what's social about it, can actually improve our self-worth. It also confirms what other studies on narcissism have shown: Genuine connection reduces the drive to feel special.

If we want to make sure we don't completely lose our center in cyberspace, we have to move from SoMe to SoWe. There are six basic strategies that can help us place genuine relationships front and center.

Surround Yourself with Real Friends

When I first entered the world of social media, I felt not just lost, but lonely. I hadn't yet connected online with any of my good friends. As a result, the digital world became little more than a blank-faced audience I needed to impress. They either liked me or they didn't—that was our only connection. And that's why I became obsessed with numbers. Our craving for admiration and attention inevitably surges when we feel disconnected.

Think about when you first meet strangers at a party. You have the pressure "to be on"—that urge to "win people over." In social media, the feeling's even worse. If you stumble over introductions or get tipsy and drop the f-bomb at a neighbor's party, you can count on people forgetting by the end of the evening and, until then, you can hide behind the cheese table. But the Internet remembers everything—forever. Your words keep floating around in full view of everyone. There's no such thing as low profile (unless you opt out). As a result, it's hard not to get caught up in making a good impression.

To relieve the pressure, do in cyberspace as you would at a party. Find people you know first, as many as you can. It's always easier to connect with strangers when you have friends in the room. And be cautious about making those new connections. Unless you're looking to build an audience professionally, collecting friends or followers in the thousands is a mistake—it's also a game narcissists often play. Without any real connections, they've no choice but to put on a show. That's

SoMe at its worst—a room full of strangers primping, preening, and posing to get attention.

Be Open

The students in the Facebook study who lingered longer than their peers over posts and images suffered a blow to their body image and overall self-esteem. Taken with other findings, the most sensible interpretation is that they were *dwelling* on how they measured up to others. That's one of the biggest dangers of social media. In real life, it's rude to stare at other peoples' bodies or to root through their houses, unearthing evidence of all the ways they're better than us. In social media you can do that all day long. But there's nothing *social* about this, and when we stop interacting with people and start gazing at their lives from a distance, we miss out on any of the rewards social media has to offer.

In some of the earliest studies on social media, the greatest benefit came to quieter, more socially withdrawn people, who found it easier to be open about the ups and downs of their lives with their online friends—new and old—than to bare their souls to people they met in their everyday lives. Writing about themselves, at their leisure, may have eliminated the pressure they felt about getting the words right in person. As a result, social media extended—even expanded—their relationships, boosting their social self-confidence and self-esteem, and making them happier about their life in general. In contrast, users who "passively" consumed information, watching posts zip by and reading about other peoples' lives, inevitably felt terrible.

In order to benefit from the time we spend on social media, we have to be open about our lives—our happy moments and sad ones, our triumphs and our traumas. And it's just as important that other people be open with us about the good and bad in their lives, too. We can't do that if we sit silently and watch our screen fill up with pictures and posts.

That's another reason to surround yourself with people you know—it gives you the chance to share more of yourself. When we accumulate too many followers or "friends," we just don't have the time to converse directly, to dig in and find out about each other, to establish a true connection—and without *that*, everyone is reduced to a smoothly polished avatar. If you surround yourself with these empty online relationships, you're bound to get sucked into a game of "whose life looks better?" because you'll have little time to do anything else as hundreds—potentially thousands—of pictures and posts scroll by. You certainly won't feel as comfortable seeking support over an especially trying day and, if you're already lonely, struggling in silence—even in cyberspace—will only make you feel worse.

But there's an even greater danger when we stop being more open about our lives in social media. When we only share selected pieces of who we are, leaving out the normal human frailties—the mistakes, the failures, the struggles of every day life—we're essentially hiding, and that's a dangerous game to play.

We never feel genuinely self-confident when we bury our true nature. We assume that whatever we're concealing is somehow shameful and that we're wise to keep it secret if we want to be liked and loved. The results, though, can be disastrous. When we fear that revealing ourselves might burden people (and drive them away), we slide into the self-abnegation of echoism; when we fear that revealing ourselves might make us seem weak or small, we often slip into the grandstanding of unhealthy narcissism. Either way, we miss out on the healthy self-esteem boost that comes from genuinely connecting with the people around us.

Find a Community with Purpose

Joining a group with a cause is a great way to feel strongly connected, and it's also a helpful way to put any desire for social media attention to healthy use. If you're feeling lonely in social media, or obsessed with likes or favorites or +1's, try participating in a forum devoted to a particular political or social cause, such as equal rights or climate-change prevention. Or follow blogs on topics that have personal relevance, such as parenting effectively or sustaining love relationships. I've seen fans show remarkable courage and openness in their comments to my and my friends' articles, and they've been amazingly caring and attentive to others in tremendous pain.

There are potentially millions of online communities that can be formed. Microblogging communities often spontaneously explode around a topic when someone adds a hashtag (#) to a posting. The results are often astonishing.

In September 2014, activist/author and domestic-abuse survivor Beverly Gooden launched the topic #whyIstayed, and posted a series of tweets describing what made it hard for her to end her relationship with her abusive husband—and how she found the strength to leave. At the time, she was responding to people's confusion over why, after football player Ray Rice had punched his fiancée, Janay Rice, in the face, she had chosen not only to remain with him but had gone on to marry him. Leaving an abuser, Gooden reminded people, is a "process not an event." She wanted to educate the world about just how hard that process can be and to empower others in the same situation to tell their own story.

Within hours, domestic-abuse survivors around the globe—many of whom are often extreme echoists—found their voice and shared their stories by adding a hashtag to their posts. The topic went viral, drawing worldwide media coverage. In the end, some survivors finally found the strength and support they needed to leave their destructive relationships. They recounted that positive journey in another thread, #whyIleft.

The empowerment created by the #whyIstayed community comes as no surprise to researchers. Decades of studies demonstrate that when people share their personal experiences within a group setting, they feel a deep sense of belonging—and often gain boosts in self-confidence, empathy, and happiness. A similar transformative effect may happen in online communities.

But not in all. Unfortunately, online communities can be toxic as well as therapeutic. Cyberspace is like the Old West—open and unregulated. And sad to say, there are highly sadistic people roaming around—aka trolls—who take great pleasure in firing off hurtful, abusive remarks, particularly to sensitive people seeking caring and understanding. Cyberspace also abounds in, you guessed it, narcissists. According to research, they're the most frequent tweeters, often bombarding their thousands of friends with foul language and provocative photos. Without monitors who can remove miscreants, sites can degenerate into shouting and invective.

So before joining an online community and baring your soul, pay close attention to how it works—who's in it and how it's run. The three ingredients of groups that have been found to be key in building greater self-awareness and assurance among members likely apply to social media communities as well:

- *Bond:* Is there an atmosphere of mutual respect and trust? Does someone curate and help correct behavior, removing people who violate the feeling of safety and trust? Are there clear guidelines—even rules—for what's expected of everyone? Are they posted so people can follow them?

- *Goal:* The purpose of the community has to be clear for it to provide the most benefit. Do you know why you're there? Is it to develop new skills? Gain information? Provide support?
- *Task:* How do people learn and grow from each other in the group? Is it by posting questions or comments? Is it by posting links? Does the community provide clear explanations for how members can contribute? The ways people connect have to be linked in some clear way to the larger goals that brought everyone together in the first place.

The #whyIstayed community, as rudimentary as it was, met these requirements. The task was clear: tell your story. The goal was obvious: raise awareness about why people have trouble leaving and end victim blaming. And the bond became powerful—posters could feel relatively safe sharing their story because followers quickly blasted trolls who tried to attack them.

If you keep these three components in mind while searching out communities, you won't just feel a greater sense of belonging in social media; you might even experience personal growth and find the strength to change your life—and the lives of others—for the better.

Avoid Image Churning

Think about how *often* you present yourself. Psychologist Christopher Carpenter, of Western Illinois University, asked 292 students to rate themselves on a series of actions. The more narcissistic people were, the greater their frequency of:

- Updating their status
- Posting new photos of themselves

- Updating their profile information
- Changing their profile photo
- Tagging themselves (labeling themselves in photos)

Any one of these behaviors turns your friends into your own personal audience. I call them image churning. They're forms of self-promotion. And if you devote too much time to them, you're taking time away from real relationships. Be leery of these behaviors in any social-media platform.

While we can't say, for sure, if they increase narcissism (not without a scientifically controlled experiment), they certainly aren't great for us. The Myspace study—which *was* a controlled experiment—found that spending time on an image-obsessed site increased people's narcissism. And all five of these activities are about image and appearance. It's a good bet that image churning *does* boost our narcissism—at least temporarily.

Be Intentional

While you're at it, *think before you post*. Feeling special because you get 40 likes for your new profile picture feels great once in a while, but it hardly nurtures a sense of closeness. I can quickly send a snapshot of myself at the beach having a great time, but has anyone really glimpsed enough of me to feel like we're connected at all? If I'm actually feeling sad because my father passed away early in the week, will a beach photo impart any sense of what I'm going through? Why did I post in the first place—was it to communicate or get attention? Will this status update bring me closer to others or leave me profoundly alone?

Take a deep breath once in a while and ask, *Why am I posting this now?* Make sure what you share does something to honor your relationships, even if it's just to give people a laugh. And while you're trying to promote healthy narcissism in yourself, think about using social media more intentionally when you're around people you love.

We can easily get lost in the brave new worlds we've created through smartphones and the online universe. We're reaching out to people we've never met who live halfway around the planet. We're finding former boyfriends and girlfriends, long-lost relatives, and forgotten classmates. We're even falling in love, thanks to the auspices of online dating sites. The possibilities are exhilaratingly endless. But it's easy to forget about who's actually standing next to us while we're caught up in all the online excitement. When we do that, we run the risk of sliding not just ourselves, but also the people we love, further up the narcissism spectrum.

Psychologist Sherry Turkle, in her TED talk "Connected but Alone," warns of the dangers of slipping mindlessly into the virtual world. It's not unusual to see adults at the theater, smartphones in hand, scrolling through messages instead of speaking to their companions, or parents sitting on playground benches, tapping away, oblivious to their children looking over to see if mom or dad is watching. We're all guilty of the occasional slip. But when it becomes a habit, we pay a steep price.

Children, especially, need to feel like their smallest accomplishments are met with rapt attention. It's why we hang even the crudest drawings by our three-year-olds on the fridge. It helps them feel special in the right way—through their parent's eyes. When social media steals attention from the people closest to us, we're foreclosing the kind of intimacy that prevents extreme narcissism.

Follow Wisely

Remember the twin fantasy—where people work hard to find all the similarities between themselves and someone else? We're all prone to "twinning" from time to time, especially when our twin is someone we admire. Half the time, we aren't even aware we're doing so. We just "become" the person.

I had a classmate years ago who loved the profanely hilarious comedian Sam Kinison. Funny in his own right, my friend didn't just parrot Kinison's best monologues, down to the exact wording and subtlest changes in inflection; he even looked like him at times—long frazzled hair and rage-filled squinty eyes. Once, when we were eating dinner together, he started screaming "Ahhhh! Ahhhhh!," Kinison-style, to punctuate his tale of a crowbar, a monkey, and a microwave (a typically surreal Kinison trifecta). The young woman we'd been sitting with clearly didn't catch the reference. She looked alarmed—then disgusted—and finally slunk away. Which was a shame, really, because I'd been working up the courage to ask her out. Later that night, I asked my friend, "Why'd you do the Kinison act? I liked that girl!"

He looked at me, genuinely confused. "I did? When?"

My friend was playing with identity. He didn't realize how completely another person's mannerisms, speech, and judgment can invade us when we emulate them. That was almost thirty years ago, long before social media or celebrity websites. You had to work hard to find videos, images, or interviews with someone you liked. Today, it's easy. Just follow them on social media.

That's fine when they model admirable qualities. Have a college age son or daughter who loves watching astrophysicist Neil deGrasse Tyson of TV's *Cosmos* expound on the wonders of the universe? Terrific. Who doesn't want a mini-Tyson milling about during school vacations? But what if they slavishly follow a conceited or vain reality TV star? Social media makes all that easier than ever. And the easiest way to boost narcissism is to emulate a narcissist.

The narcissistic thrill we get from social media needn't be bad for us at all if we follow some simple guidelines. The rules that apply to real life also apply to our digital lives. What should be clear by now is that secure love and caring relationships are the greatest protection we have from unhealthy narcissism, whether your connections are live and in person, or virtual and in cyberspace.

12

—

A PASSIONATE LIFE

THE ULTIMATE GIFT
OF HEALTHY NARCISSISM

> Great ambition is the passion of a great character. Those endowed with it may perform very good or very bad acts. All depends on the principles which direct them. —NAPOLEON BONAPARTE

> Our true passions are selfish. —STENDHAL

Picture yourself at your most physically passionate: in the arms of someone you love, touching and kissing, your mind racing with images of what comes next. Or imagine expressing your passion in another way. Maybe it's cooking a delicious meal, putting together a snappy new outfit, or reading a racy romance novel. You'll notice something about each of these moments. For a time—seconds, minutes, hours—you're not focused on what other people want or need; instead, you're lost in your own imagination, your own desires, and seeing where they lead you. Time may seem to stand still. When we're truly excited, we privilege our desires above everyone and everything else.

Intense passion, in other words, is always a little narcissistic. We can't build or create or explore in our own lives with any kind of joy, when we're fixated on what other people want. At the same time, we have to be careful in our quest for a more thrilling life. Passion has to be balanced with caring and concern for others. When we don't take other people into account, passion becomes empty, or worse, destructive.

Ironically, Napoleon, the self-proclaimed emperor who ruthlessly carved a realm across Europe, seemed to understand this, even if he failed to apply it to his own actions. We can point to many examples of passionate people throughout history who visited great evil upon the world: Hitler, Stalin, Saddam Hussein. Narcissists in our everyday lives, especially the gregarious ones, often have plenty of passion because they give free rein to their needs and desires. That's why we often find so many of them attractive. Nothing stands in the way of what they want, whether they're seducing a date, tackling risky business ventures, or defeating a rival sports team. But their passion, in the end, is neither genuine nor fulfilling because even our deepest desires become tenuous if we lose connection to the people we love. British psychoanalyst and child therapist Donald Winnicott poignantly captured the truth of this in his studies of children lost in creative play.

Picture a girl around one and a half years old, picking up blocks and dolls and stuffed animals and stacking them to form lopsided towers. At this age, she's still a little nervous to be alone. Out of the corner of her eye, she's continually checking to make sure you're there. But she goes on playing, fascinated by her discoveries. What allows our little builder to so immerse herself is that when she looks up, *you're there*. She can stack and stack and stack because she feels your presence. The world is hers, to do with as she pleases, thanks to your love. She senses that she, and what she is doing, is important—special—to you. And that allows her find her passion.

But what if you look bored or rearrange her sculptures according to your own vision? What if you leave? The conclusion her mind comes to is that what she's doing doesn't matter. Worse, she'll worry that if she surrenders to her imagination, forgetting about you for an instant, maybe you won't stick around.

Maybe she'll stop playing altogether, obsessed with regaining your interest. She might put aside the blocks and dolls and tug at your shirt or cry. Or she might, in anger, shut you out altogether, deciding the only possible way she can follow her own desire, and build anything at all, is if she *never* acknowledges your presence.

The little girl loses something vital either way. Either she sacrifices her passion to keep your interest or she sacrifices love to keep her passion. As a result, her desires can never be authentic; they're either muted or amplified.

Echoists, like the little girl who stops playing to make sure you stay with her, suppress their desires, trading their passion for deep concern; they're far too afraid of hurting others to risk seeing where their imagination leads them. They have an urge to do what feels good in the moment and then simply let it go, never letting it become so important that all other considerations fade into the background. Their passion, as a result, becomes too fragile for them build a life guided by their own imagination.

Sandy, the birthday-loathing administrative assistant we met in Chapter 3, slogged through work, joylessly overperforming in every task, not to impress, but to send a message: "I'm fine. Don't worry about me." She overachieved as an adult for the same reason she did as a child—to unburden those around her. But she never felt a thrill, even when she did well, because none of her efforts, ultimately, felt like they were for her. She had little life outside work, in part because she spent all her time making sure her boyfriend, Joe, had everything he needed. "There's too much to do!" she'd exclaim. "Who has time to loaf around having fun!"

Narcissists, in contrast, take after the little girl who becomes enraged by constant interference or absence; they trade the capacity to love for blind passion. Some have tons of confidence in bed, which can be a turn on for a while, but the fact that they won't engage emotionally eventually torpedoes the romance. Their desire is fragile in a different way; it's too frenzied and pressured, as if they can only preserve it by closing their eyes and shutting you out. In their minds, your very presence threatens their efforts to imagine—and build—the life they want.

Gary, the cocky undergrad from Chapter 3, was charming enough to attract lots of women, but he had only brief flings. His love life was predictable, by his own admission. "I lose interest once we've slept together," he said. He grew nervous that a woman would "take over," and interfere in his grand scheme to become a millionaire by age 25. One gets the feeling, with many narcissists, that they can't hold on to their enthusiasm for anything if people get too close to them. As if, like a reflection in water, their most thrilling dreams are easily dispersed by the slightest touch.

In contrast, people with healthy narcissism, like the little girl passionately lost in play, freely follow their desires precisely because they feel *special to* the people they love—and it exhilarates them, filling them with creative enthusiasm for whatever they do. Their passion, compared to the narcissist's, is made of sturdier stuff; they can hold on to their grand dreams and thirst for life without worrying love might get in the way.

Lisa, the outgoing activist we met in Chapter 3, frequently

enjoyed her favorite hobbies, bike racing and wine tasting, by herself, but she was just as content to have her husband, Doug, join her when he could. Rather than threatening her plans and dreams, his presence gave her more energy to bring them to fruition and she took great delight in sharing them. "I want us to build an amazing life *together*," she observed during one of our meetings. At the same time, Lisa derived almost as much pleasure from supporting Doug's interests. "I love helping him plan his weekends hiking in the mountains by himself. That's *his* passion. And he always comes back looking happy!"

Healthy narcissism unlocks authentic passion—the kind that emerges from within, never growing destructive or slipping away—only when we're capable of secure, loving relationships. We can follow our desire, see where it takes us, because we know—or rather, *trust*—that when we look up, the people we care about will still be there. The drive to feel special is the wellspring of passion, but love keeps the passion pure.

If authentic passion is our reward for allowing ourselves to feel special to those we love, what then do the people around us gain? The answer seems to be genuine intimacy.

The word *intimate* can be traced to ancient Latin and Indo-European words meaning *knowledge* or *to know*. Intimacy has everything to do with knowledge. No one can truly be close to us if we don't know ourselves because we can't share who we are. Echoists spend so little time exploring their inner life that they're completely unfamiliar with it. But narcissists, as you've seen, are just as much in the dark, preferring an elaborate and often fanciful idea of themselves. Their denial about their flaws and failings reaches such depths that the persona they present—Winnicott called it a "false self"—barely resembles a real human being. And introverted narcissists are so terrified of exposing their weaknesses, both real and imagined, that they rarely allow people close enough to see what they really need.

When people feel important enough to pay special attention to their deepest desires and needs—and honestly share them—those who care about them learn something new. They finally get to meet the person they love, a truly thrilling moment for all involved. It's a privilege that even narcissists and echoists can enjoy once they move closer to the center of the spectrum, as I learned when Jean, the depressed, empty nester you met in Chapter 5, brought her husband, Robert, to talk about their recent progress together.

Jean had begun more and more to allow herself a special place in her husband's life, using empathy prompts whenever possible. "I told him he's the most important person to me," Jean said, sniffling and glancing at her husband. "And that I'm so afraid of losing him, of not being able to enjoy the remaining years we have together. I just want to spend them with *him*."

"Had you ever heard Jean feel so sad about losing you before?" I asked Robert.

"Never," he said, his face softening as he looked at Jean. "In the past she'd mostly given me the cold shoulder. In fact, I don't think anyone's ever said they *need* me, let alone miss me, like Jean did."

Jean smiled. "He held me when I said that and I cried."

"And you, Robert?" I asked.

"Something changed that day," he replied. "Something in me was different. I felt like I mattered to someone, that I could relax a little." This was big for him; in the past, he'd chased after one special high after another, never feeling like he'd won enough people over with his charm or good looks. "My own mother only ever used me to show off," he said, sadly, "telling people how handsome I was or how smart, but I can't recall a time I felt like she really saw me for who I was. Not like Jean has. I found myself telling her things I've *never* told anyone before."

And Robert had discovered more about Jean, too. She'd begun taking swing-dancing lessons, twice a week, and surprised Robert one night by inviting him out on a date.

"I didn't even know she could dance!" Robert said, laughing. "Now we take lessons together."

Robert and Jean now felt *special to* each other, and it made each day a new adventure. They'd come to truly know one another. By finally shifting to the center of the spectrum, Jean had given Robert the gift of true intimacy, and he'd returned the favor.

It turns out that even Echo and Narcissus make a pretty good pair, once they meet in the middle.

We're all a bit like Narcissus. We tramp through the forest of life and we meet people along the way, each with their own talents, their own desires, their own need to feel special. Just think how different the young man's life would have been if rather than ignoring those he met and disappearing into himself, he'd stayed to chat, share a meal, and ask them about themselves, and then continued on his way. Just think how different Echo's life might have been had she stopped to glance at her reflection in the enchanted pool, taken a little delight at her image, and decided to dive in and reemerge, feeling invigorated by her journey into herself. Perhaps she might have broken her own curse and found her voice again.

A good life balances our own self-interests with others people's needs. That's healthy narcissism. It's what gives us the energy to build a life full of adventure and self-discovery. Healthy narcissism is where passion and compassion merge, offering a truly exhilarating life. And that's a pretty great place to be.

ACKNOWLEDGMENTS

Many of the concepts in this book were inspired by the remarkable training experiences and gifted supervisors I've had over the years; each one of them helped me become a better clinician. Special thanks to Drs. Joe Shay, Kenneth Zack, and June Wolf for all their wisdom and support over the years.

Two men shaped my career long after I left my internship and I miss them dearly: my former supervisor, Dr. Andy Morrison, authored and edited several books on narcissism and devoted his life to understanding the subject; and Dr. Gerry Kochansky, who, likewise, had a gift for reaching extremely narcissistic and borderline clients. I often thought of them while writing this book.

I also want to thank my clients, who not only granted me the privilege of hearing their stories, but also trusted me to help; without their courage, I couldn't have arrived at the insights and lessons in this book.

It's often said that writing a book is a team effort, and I feel fortunate to have found a home with the immensely talented group of people at HarperCollins. Thank you to Carole

Tonkinson, the UK HarperCollins editor who approached me about publishing a book on narcissism in the first place; she introduced me to the entire HarperCollins team, including my wonderful editors at HarperWave. Karen Rinaldi brought both insight and energy to this entire project, from conception to completion. Julie Will's keen editorial eye, incisive comments, and knowledge of psychology made the final process of refining *Rethinking Narcissism* an absolute pleasure.

My publicity and marketing team—Tina Andreadis, Kate D'Esmond, and Stephanie Cooper—demonstrated not only tremendous skill but also caring and attentiveness in their efforts to make this book a success.

I also could never have developed a new measure, if my colleague, Professor Stuart Quirk, of Central Michigan University, hadn't shared my enthusiasm for bringing some empirical rigor to the assessment of healthy narcissism. I feel blessed to have met him during our training together and I look forward to continuing our work on the Narcissism Spectrum Scale.

I'm indebted, too, to my insatiably curious and fearlessly supportive agent, Miriam Altshuler. Many years ago, one crisp fall evening, she called me up with the thrilling news that she wanted to represent me; I'm forever delighted that she did.

I'm eternally grateful to my editor and friend, Anastasia Toufexis, who helped shape my voice and sharpen my thinking. Thanks to her continual warmth, much-needed humor, exquisite ear for language, and remarkable gift for logic, I've become a far better writer.

My heartfelt thanks go to Lisa Tener, book coach and friend, who helped me find my passion as an author.

Thank you as well, to *The Huffington Post* and *Psychology Today* for giving me a forum to share my ideas.

My brother, Brian Malkin, and the entire team of lawyers at Ference & Associates, worked hard to obtain the permissions for the manuscript. I'd like to thank them for their tireless efforts.

Finally, special thanks to my family and friends who good-naturedly listened to the trials and travails of this process, read and commented on early drafts, and provided me with entertaining distractions when I needed them. Thank you for all the scintillating and encouraging conversations; you reminded me that no matter what I do—succeed or fail—I've got a group of people who think I'm special—and I'm happy to say, the feelings are mutual.

And of course, thank you to my amazing wife, Jennifer, and my daughters, Anya and Devin, without whose continued support, both emotional and practical, I couldn't possibly have written a single article, let alone an entire book.

RESOURCES

The strategies in Chapter 8 owe much to the work of Dr. Sue Johnson, inventor of Emotionally Focused Therapy (EFT) for couples. The cutting-edge research on narcissists, reviewed throughout *Rethinking Narcissism*, confirms what Dr. Johnson has found in her groundbreaking work on couples—in any relationship, our best chance at change lies in going beneath our anger and withdrawal, to share our deeper needs and feelings.

Through EFT, I've seen countless couples, in workshops and in my own practice, break through any number of bad habits—including unhealthy narcissism—and look to each other for support and appreciation in ways that increase their intimacy (her approach has an 86 percent success rate, a figure that's unparalleled in psychotherapy). If you think you or your partner are struggling with unhealthy narcissism, I highly recommend Dr. Sue Johnson's books:

> *Hold Me Tight: Seven Conversations for a Lifetime of Love* (Little, Brown, 2008)
> *Love Sense: The Revolutionary New Science of Romantic Relationships* (Little, Brown, 2013)

If you're looking for more information about workshops and groups, or want to find a therapist, you can search for local certified EFT therapists at www.iceeft.com.

Dr. John Gottman has also done much to illuminate the field of couples therapy. Along with his wife, Julie Gottman, also a therapist, he travels the world educating couples on how to break their toxic patterns of interaction and restore intimacy. Like Dr. Johnson, he's been widely published in scientific journals for his work; his research through his Love Lab—watching countless hours of couples interacting, live and on video—has yielded some of the most important ingredients for making a marriage last.

If you want to improve your relationship, you might enjoy:

10 Lessons to Transform Your Marriage: America's Love Lab Experts Share Their Strategies for Strengthening Your Relationship (Random House, 2007)

The Seven Principles for Making Marriage Work: A Practical Guide (Random House, 2000)

If you'd like to find a Gottman-certified therapist, see www.gottman .com.

Dr. Diana Fosha's Accelerated Experiential Dynamic Psychotherapy (AEDP) has revolutionized how we understand growth and change in psychotherapy. Combining neurobiology, attachment theory, emotion theory, affective neuroscience, body-focused approaches, and interpersonal neurobiology, AEDP effects deeper change by helping people experience themselves, the world, and their relationships in a brand-new way; in doing so, it opens them up to profound and lasting transformation.

To learn more about AEDP, see Dr. Fosha's textbook, The Transforming Power of Affect: A Model for Accelerated Change (Basic Books, 2000).

To find an AEDP certified therapist, visit https://www.aedpinstitute .org/find-an-aedp-institute-therapist/.

Finally, Dr. Les Greenberg has transformed the way we think about change in individual psychotherapy. If you think you're a narcissist and want to change, you'll have to dig deep and become comfortable with feelings you may not even have realized you had. Dr. Greenberg's Emotion Focused Therapy helps people do just that.

You can find therapists and resources for individual EFT at http:// www.emotionfocusedclinic.org/efttrained.htm.

REFERENCES

1. Rethinking Narcissism: Old Assumptions, New Ideas

Ackerman, R. A. *The Role of Narcissism in Romantic Relationship Initiation.* ProQuest Information & Learning, 2012, dissertation 72.

Ackerman, R. A., E. A. Witt, M. B. Donnellan, K. H. Trzesniewski, R. W. Robins, and D. A. Kashy. What does the narcissistic personality inventory really measure? *Assessment*, 2011, vol. 18(1), pp. 67–87.

Alicke, M. D., and C. Sedikides. *Handbook of Self-Enhancement and Self-Protection.* Guilford Press, 2011.

Barelds, D. P., and P. Dijkstra. Positive illusions about a partner's physical attractiveness and relationship quality. *Personal Relationships*, 2009, vol. 16(2), pp. 263–83.

Baumeister, R. F., and K. D. Vohs. Narcissism as addiction to esteem. *Psychological Inquiry*, 2001, vol. 12(4), pp. 206–10.

Bonanno, G. A., N. P. Field, A. Kovacevic, and S. Kaltman. Self-enhancement as a buffer against extreme adversity: Civil war in Bosnia and traumatic loss in the United States. *Personality and Social Psychology Bulletin*, 2002, vol. 28(2), pp. 184–96.

Brown, J. D. Across the (not so) great divide: Cultural similarities in self-evaluative processes. *Social and Personality Psychology Compass*, 2010, vol. 4(5), pp. 318–30.

Brown, J. D. Understanding the better than average effect: Motives (still) matter. *Personality and Social Psychology Bulletin*, 2012, vol. 38(2), pp. 209–19.

Furnham, A., D. J. Hughes, and E. Marshall. Creativity, OCD, Narcissism and the Big Five. *Thinking Skills and Creativity*, 2013, vol. 10, pp. 91–98.

Goorin, L., and G. A. Bonanno. Would you buy a used car from a self-enhancer? Social benefits and illusions in trait self-enhancement. *Self and Identity*, 2009, vol. 8(2,3), pp. 162–75.

Jakobwitz, S., and V. Egan. The dark triad and normal personality traits. *Personality and Individual Differences*, 2006, vol. 40(2), pp. 331–39.

Konrath, S., B. P. Meier, and B. J. Bushman. Development and Validation of the Single Item Narcissism Scale (SINS). *PLoS One*, 2014, vol. 9(8), e103469.

Le, B., N. L. Dove, C. R. Agnew, M. S. Korn, and A. A. Mutso. Predicting nonmarital romantic relationship dissolution: A meta-analytic synthesis. *Personal Relationships*, 2010, vol. 17(3), pp. 377–90.

Murray, S. L., J. G. Holmes, and D. W. Griffin. The benefits of positive illusions: Idealization and the construction of satisfaction in close relationships. *Journal of Personality and Social Psychology*, 1996, vol. 70(1), p. 79.

O'Mara, E. M., L. Gaertner, C. Sedikides, X. Zhou, and Y. Liu. A longitudinal-experimental test of the panculturality of self-enhancement: Self-enhancement promotes psychological well-being both in the west and the east. *Journal of Research in Personality*, 2012, vol. 46(2), pp. 157–63.

Pincus, A. L. The Pathological Narcissism Inventory. In *Understanding and Treating Pathological Narcissism*, J. Ogrudniczuk, editor, pp. 93–110. American Psychological Association, 2013.

Taylor, S. E., J. S. Lerner, D. K. Sherman, R. M. Sage, and N. K. McDowell. Are self-enhancing cognitions associated with healthy or unhealthy biological profiles? *Journal of Personality and Social Psychology*, 2003, vol. 85(4), pp. 605–15.

Taylor, S. E., J. S. Lerner, D. K. Sherman, R. M. Sage, and N. K. McDowell. Portrait of the self-enhancer: Well adjusted and well liked or maladjusted and friendless? *Journal of Personality and Social Psychology*, 2003, vol. 84(1), p. 165.

2. Confusion and Controversy: How Narcissism Became a Dirty Word and We Found an Epidemic

Annas, J. Self-love in Aristotle. *The Southern Journal of Philosophy*, 1989, vol. 27(S1), pp. 1–18.

Arnett, J. J. The evidence for Generation We and against Generation Me. *Emerging Adulthood*, 2013, vol. 1(1), pp. 5–10.

Back, M. D., A. C. Kufner, M. Dufner, T. M. Gerlach, J. F. Rauthmann, and J. J. Denissen. Narcissistic admiration and rivalry: Disentangling the bright and dark sides of narcissism. *Journal of Personality and Social Psychology*, 2013, vol. 105(6), pp. 1013–37.

Cramer, P. Freshman to senior year: A follow-up study of identity, narcissism, and defense mechanisms. *Journal of Research in Personality*, 1998, vol. 32(2), pp. 156–72.

Denuy, D. J. Self-love and benevolence. *Reason Papers*, 1983, vol. 9, pp. 57–60.

Donnellan, M. B., K. H. Trzesniewski, and R. W. Robins. An emerging epidemic of narcissism or much ado about nothing? *Journal of Research in Personality*, 2009, vol. 43(3), pp. 498–501.

Dufner, M., J. J. Denissen, M. van Zalk, B. Matthes, W. H. Meeus, M. A. van Aken, and C. Sedikides. Positive intelligence illusions: on the relation between intellectual self-enhancement and psychological adjustment. *Journal of Personality*, 2012, vol. 80(3), pp. 537–71.

Grijalva, E., P. D. Harms, D. A. Newman, B. H. Gaddis, and R. C. Fraley. Narcissism and leadership: A meta-analytic review of linear and nonlinear relationships. *Personnel Psychology*, 2015, in press.

Hill, P. L., and Lapsley, D. K. Adaptive and maladaptive narcissism in adolescent development. In *Narcissism and Machiavellianism in Youth: Implications for the development of adaptive and maladaptive behavior*, C. T. Barry, P. K. Kerig, K. K. Stellwagen, T. D. Barry, editors, pp. 89–105. American Psychological Association, 2011.

Kenny, M. *Narcissistic illusions: An empirical typology.* Dissertation Abstracts International: Section B: The Sciences and Engineering, 2001, vol. 62(10-B) p. 4819.

Kreyche, J. How we are moral: Benevolence, utility, and self-love in Hobbes and Hume. *Stance*, 2011, vol. 4, p. 27.

Lapsley, D. K., and M. C. Aalsma. An empirical typology of narcissism and mental health in late adolescence. *Journal of Adolescence*, 2006, vol. 29(1), pp. 53–71.

Lasch, C. *The Culture of Narcissism: American life in an age of diminishing expectations.* W. W. Norton & Company, 1991.

Lunbeck, E. *The Americanization of Narcissism.* Harvard University Press, 2014.

Millennials: A Portrait of Generation Next—Confident. Connected. Open to Change. P. Taylor and S. Keeter, editors. Pew Research Center, 2010. http://www.pewsocialtrends.org/files/2010/10/millennials-confident-connected-open-to-change.pdf.

Miller, J. D., J. McCain, D. R. Lynam, L. R. Few, B. Gentile, J. MacKillop, and W. K. Campbell. A comparison of the criterion validity of popular measures of narcissism and narcissistic personality disorder via the use of expert ratings. *Psychological Assessment*, 2014, vol. 26(3), pp. 958–69.

Miller, J. D., J. Price, and W. K. Campbell. Is the Narcissistic Personality Inventory still relevant? A test of independent grandiosity and entitlement scales in the assessment of narcissism. *Assessment*, 2012, vol. 19(1), pp. 8–13.

Mitchell, S. A. *Relational Concepts in Psychoanalysis: An integration.* Harvard University Press, 1988.

Paunonen, S. V., J.-E. Lönnqvist, M. Verkasalo, S. Leikas, and V. Nissinen. Narcissism and emergent leadership in military cadets. *The Leadership Quarterly*, 2006, vol. 17(5), pp. 475–86.

Raskin, R. N. and C. S. Hall. A narcissistic personality inventory. *Psychological Reports*, 1979, vol. 45(2), p. 590.

Roberts, B. W., G. Edmonds, and E. Grijalva. It is developmental me, not generation me: Developmental changes are more important than generational changes in narcissism—Commentary on Trzesniewski & Donnellan (2010). *Perspectives on Psychological Science*, 2010, vol. 5(1), pp. 97–102.

Ronningstam, E. *Disorders of Narcissism: Diagnostic, clinical, and empirical implications.* American Psychiatric Press, 1998.

Rosenthal, S. A., and J. M. Hooley. Narcissism assessment in social-personality research: Does the association between narcissism and psychological health result from a confound with self-esteem? *Journal of Research in Personality*, 2010, vol. 44(4), pp. 453–65.

Rosenthal, S. A., R. Matthew Montoya, L. E. Ridings, S. M. Rieck, and J. M. Hooley. Further evidence of the Narcissistic Personality Inventory's validity problems: A meta-analytic investigation—Response to Miller, Maples, and Campbell (this issue). *Journal of Research in Personality*, 2011, vol. 45(5), pp. 408–16.

Sosik, J., J. Chun, and W. Zhu. Hang on to your ego: The moderating role of leader narcissism on relationships between leader charisma and follower psychological empowerment and moral identity. *Journal of Business Ethics*, 2014, vol. 120(1), pp. 65–80.

Strozier, C. B. *Heinz Kohut: The making of a psychoanalyst.* Farrar, Straus and Giroux, 2001.

Taylor, S. E., J. S. Lerner, D. K. Sherman, R. M. Sage, and N. K. McDowell. Are self-enhancing cognitions associated with healthy or unhealthy biological profiles? *Journal of Personality and Social Psychology*, 2003, vol. 85(4), pp. 605–15.

Trull, T. J. Ruminations on narcissistic personality disorder. *Personality Disorders: Theory Research and Treatment*, 2014, vol. 5(2), pp. 230–31.

Trzesniewski, K. H., and M. B. Donnellan. Reevaluating the evidence for increasingly positive self-views among high school students: More evidence for consistency across generations (1976–2006). *Psychological Science*, 2009, vol. 20(7), pp. 920–22.

Trzesniewski, K. H., and M. B. Donnellan. Rethinking "Generation Me": A study of cohort effects from 1976–2006. *Perspectives on Psychological Science*, 2010, vol. 5(1), pp. 58–75.

Trzesniewski, K. H., M. B. Donnellan, and R. W. Robins. Do today's young people really think they are so extraordinary? An examination of secular trends in narcissism and self-enhancement. *Psychological Science*, 2008, vol. 19(2), pp. 181–88.

Twenge, J. M. *Generation Me: Why today's young Americans are more confident, assertive, entitled—and more miserable than ever before.* Free Press, 2006.

Twenge, J. M., and W. K. Campbell. *The Narcissism Epidemic: Living in the age of entitlement.* Free Press, 2009.

3. From 0 to 10: Understanding the Spectrum

Back, M. D., A. C. Kufner, M. Dufner, T. M. Gerlach, J. F. Rauthmann, and J. J. Denissen. Narcissistic admiration and rivalry: Disentangling the bright and dark sides of narcissism. *Journal of Personality and Social Psychology*, 2013, vol. 105(6), pp. 1013–37.

Campbell, W. K., and S. M. Campbell. On the self-regulatory dynamics created by the peculiar benefits and costs of narcissism: A contextual reinforcement model and examination of leadership. *Self and Identity*, 2009, vol. 8(2,3), pp. 214–32.

Cramer, P. Narcissism through the ages: What happens when narcissists grow older? *Journal of Research in Personality*, 2011, vol. 45(5), pp. 479–92.

Deluga, R. J. Relationship among American presidential charismatic leadership, narcissism, and rated performance. *The Leadership Quarterly*, 1997, vol. 8(1), pp. 49–65.

Dickinson, K. A., and A. L. Pincus. Interpersonal analysis of grandiose and vulnerable narcissism. *Journal of Personality Disorders*, 2003, vol. 17(3), pp. 188–207.

Dufner, M., J. J. Denissen, M. van Zalk, B. Matthes, W. H. Meeus, M. A. van Aken, and C. Sedikides. Positive intelligence illusions: On the relation between intellectual self-enhancement and psychological adjustment. *Journal of Personality*, 2012, vol. 80(3), pp. 537–71.

Edelstein, R. S., N. J. Newton, and A. J. Stewart. Narcissism in midlife: Longitudinal changes in and correlates of women's narcissistic personality traits. *Journal of Personality*, 2012, vol. 80(5), pp. 1179–1204.

Foster, J. D., W. Keith Campbell, and J. M. Twenge. Individual differences in narcissism: Inflated self-views across the lifespan and around the world. *Journal of Research in Personality*, 2003, vol. 37(6), pp. 469–86.

Gebauer, J. E., C. Sedikides, B. Verplanken, and G. R. Maio. Communal narcissism. *Journal of Personality and Social Psychology*, 2012, vol. 103(5), pp. 854–78.

Gebauer, J. E., J. Wagner, C. Sedikides, and W. Neberich. Agency-communion and self-esteem relations are moderated by culture, religiosity, age, and sex: Evidence for the "self-centrality breeds self-enhancement" principle. *Journal of Personality*, 2013, vol. 81(3), pp. 261–75.

Grijalva, E., D. Newman, L. Tay, M. B. Donnellan, P. Harms, R. Robins, and T. Yan Gender Differences in narcissism: A meta-analytic review. *Psychological Bulletin*, 2015, in press.

Helson, R., V. S. Y. Kwan, O. P. John, and C. Jones. The growing evidence for personality change in adulthood: Findings from research with personality inventories. *Journal of Research in Personality*, 2002, vol. 36(4), pp. 287–306.

Hendin, H. M., and J. M. Cheek. Assessing hypersensitive narcissism: A reexamination of Murray's Narcism Scale. *Journal of Research in Personality*, 1997, vol. 31(4), pp. 588–99.

Hill, P. L., and B. W. Roberts. Narcissism, well-being, and observer-rated personality across the lifespan. *Social Psychological and Personality Science*, 2012, vol. 3(2), pp. 216–23.

Hill, R. W., and G. P. Yousey. Adaptive and maladaptive narcissism among university faculty, clergy, politicians, and librarians. *Current Psychology*, 1998, vol. 17(2, 3), pp. 163–69.

Luo, Y. L., H. Cai, C. Sedikides, and H. Song. Distinguishing communal narcissism from agentic narcissism: A behavior genetics analysis on the agency-communion model of narcissism. *Journal of Research in Personality*, 2014, vol. 49, pp. 52–58.

Mark Young, S., and D. Pinsky. Narcissism and celebrity. *Journal of Research in Personality*, 2006, vol. 40(5), pp. 463–71.

Mathieu, C., and E. St-Jean. Entrepreneurial personality: The role of narcissism. *Personality and Individual Differences*, 2013, vol. 55(5), pp. 527–31.

Roberts, B. W., D. Wood, and J. L. Smith. Evaluating five factor theory and social investment perspectives on personality trait development. *Journal of Research in Personality*, 2005, vol. 39(1), pp. 166–84.

Taylor, S. E., J. S. Lerner, D. K. Sherman, R. M. Sage, and N. K. McDowell. Portrait of the self-enhancer: Well adjusted and well liked or maladjusted and friendless? *Journal of Personality and Social Psychology*, 2003, vol. 84(1), p. 165.

Wink, P. Three types of narcissism in women from college to mid-life. *Journal of Personality*, 1992, vol. 60(1), pp. 7–30.

4. The Narcissism Test: How Narcissistic Are You?

Malkin, C., and Quirk, S. Evidence for the reliability and construct validity of the Narcissism Spectrum Scale. *Research in progress,* www.chsbs.cmich.edu/NSS.

Raskin, R. N., and C. S. Hall. A narcissistic personality inventory. *Psychological Reports,* 1979, vol. 45(2), p. 590.

5. Root Causes: The Making of Echoists and Narcissists

Andersen, S. M., R. Miranda, and T. Edwards. When self-enhancement knows no bounds: Are past relationships with significant others at the heart of narcissism? *Psychological Inquiry,* 2001, vol. 12 (4), pp. 197–202.

Barry, C. T., P. J. Frick, K. K. Adler, and S. J. Grafeman. The predictive utility of narcissism among children and adolescents: Evidence for a distinction between adaptive and maladaptive narcissism. *Journal of Child and Family Studies,* 2007, vol. 16(4), pp. 508–21.

Bosson, J. K., C. E. Lakey, W. K. Campbell, V. Zeigler-Hill, C. H. Jordan, and M. H. Kernis. Untangling the links between narcissism and self-esteem: A theoretical and empirical review. *Social and Personality Psychology Compass,* 2008, vol. 2(3), pp. 1415–39.

Campbell, W. K., C. P. Bush, A. B. Brunell, and J. Shelton. Understanding the social costs of narcissism: The case of the tragedy of the commons. *Personality and Social Psychology Bulletin,* 2005, vol. 31(10), pp. 1358–68.

Campbell, W. K., and J. D. Miller. *The Handbook of Narcissism and Narcissistic Personality Disorder: Theoretical approaches, empirical findings, and treatments.* John Wiley & Sons, 2011.

Cater, T. E., V. Zeigler-Hill, and J. Vonk. Narcissism and recollections of early life experiences. *Personality and Individual Differences,* 2011, vol. 51(8), pp. 935–39.

Cramer, P. Young. Adult Narcissism: A 20 year longitudinal study of the contribution of parenting styles, preschool precursors of narcissism, and denial. *Journal of Research in Personality,* 2011, vol. 45(1), pp. 19–28.

Cramer, P., and C. J. Jones. Narcissism, identification, and longitudinal change in psychological health: Dynamic predictions. *Journal of Research in Personality,* 2008, vol. 42(5), pp. 1148–59.

Ettensohn, M. D. *The Relational Roots of Narcissism: Exploring relationships between attachment style, acceptance by parents and peers, and measures of grandiose and vulnerable narcissism.* Dissertation Abstracts International: Section B: The Sciences and Engineering, 2013 vol. 73(10-B)(E).

Horton, R. S. On environmental sources of child narcissism: Are parents really to blame. In *Narcissism and Machiavellianism in Youth: Implications for the development of adaptive and maladaptive behavior,* C. T. Barry, P. K. Kerig, K. K. Stellwagen, T. D. Barry, editors, pp. 125–43. American Psychological Association, 2011.

Jakobwitz, S.. and V. Egan. The dark triad and normal personality traits. *Personality and Individual Differences,* 2006, vol. 40(2), pp. 331–39.

Morf, C. C., and F. Rhodewalt. Expanding the dynamic self-regulatory processing model of narcissism: Research directions for the future. *Psychological Inquiry,* 2001, vol. 12(4), pp. 243–51.

Morrison, A. P. *Shame: The underside of narcissism.* Analytic Press, 1989.

Myers, E. M., and V. Zeigler-Hill. How much do narcissists really like themselves? Using the bogus pipeline procedure to better understand the self-esteem of narcissists. *Journal of Research in Personality,* 2012, vol. 46(1), pp. 102–5.

Otway, L. J., and V. L. Vignoles. Narcissism and childhood recollections: A quantitative test of psychoanalytic predictions. *Personality and Social Psychology Bulletin,* 2001, vol. 32(1), pp. 104–16.

Rappoport, A. Co-narcissism: How we accommodate to narcissistic parents, 2005, http://www.alanrappoport.com.

Rohmann, E., E. Neumann, M. J. Herner, and H.-W. Bierhoff. Grandiose and vulnerable narcissism. *European Psychologist,* 2012, vol. 17(4), pp. 279–90.

Segrin, C., A. Woszidlo, M. Givertz, and N. Montgomery. Parent and child traits associated with overparenting. *Journal of Social and Clinical Psychology,* 2013, vol. 32(6), pp. 569–95.

Smolewska, K., and K. Dion. Narcissism and adult attachment: A multivariate approach. *Self and Identity,* 2005, vol. 4(1), pp. 59–68.

Tolmacz, R., and M. Mikulincer. The sense of entitlement in romantic relationships—Scale construction, factor structure, construct validity, and its associations with attachment orientations. *Psychoanalytic Psychology,* 2011, vol. 28(1), p. 75.

Trumpeter, N. N., P. Watson, B. J. O'Leary, and B. L. Weathington. Self-functioning and perceived parenting: Relations of parental empathy and love inconsistency with narcissism, depression, and self-esteem. *The Journal of Genetic Psychology,* 2008, vol. 169(1), pp. 51–71.

Vernon, P. A., V. C. Villani, L. C. Vickers, and J. A. Harris. A behavioral genetic investigation of the Dark Triad and the Big 5. *Personality and Individual Differences,* 2008, vol. 44(2), pp. 445–52.

Watson, P., S. E. Hickman, R. J. Morris, J. T. Milliron, and L. Whiting. Narcissism, self-esteem, and parental nurturance. *The Journal of Psychology,* 1995, vol. 129(1), pp. 61–73.

Zeigler-Hill, V., and A. Besser. A glimpse behind the mask: Facets of narcissism and feelings of self-worth. *Journal of Personality Assessment*, 2013, vol. 95(3), pp. 249–60.

Zeigler-Hill, V., B. A. Green, R. C. Arnau, T. B. Sisemore, and E. M. Myers. Trouble ahead, trouble behind: Narcissism and early maladaptive schemas. *Journal of Behavior Therapy and Experimental Psychiatry*, 2011, vol. 42(1), pp. 96–103.

6. Echoism and Narcissism: From Bad to Worse

Ackerman, R. A., and M. B. Donnellan. Evaluating self-report measures of narcissistic entitlement. *Journal of Psychopathology and Behavioral Assessment*, 2013, vol. 35(4), pp. 460–74.

Andersen, S. M., R. Miranda, and T. Edwards. When self-enhancement knows no bounds: Are past relationships with significant others at the heart of narcissism? *Psychological Inquiry*, 2001, vol. 12(4), pp. 197–202.

Cheng, J. T., J. L. Tracy, and G. E. Miller. Are narcissists hardy or vulnerable? The role of narcissism in the production of stress-related biomarkers in response to emotional distress. *Emotion*, 2013, vol. 13(6), pp. 1004–11.

Foster, J. D., I. Shrira, and W. K. Campbell. Theoretical models of narcissism, sexuality, and relationship commitment. *Journal of Social and Personal Relationships*, 2006, vol. 23(3), pp. 367–86.

Holtzman, N. S., S. Vazire, and M. R. Mehl. Sounds like a narcissist: Behavioral manifestations of narcissism in everyday life. *Journal of Research in Personality*, 2010, vol. 44(4), pp. 478–84.

Horvath, S., and C. C. Morf. To be grandiose or not to be worthless: Different routes to self-enhancement for narcissism and self-esteem. *Journal of Research in Personality*, 2010, vol. 44(5), pp. 585–92.

Miller, J. D., W. K. Campbell and P. A. Pilkonis. Narcissistic personality disorder: Relations with distress and functional impairment. *Comprehensive Psychiatry*, 2007, vol. 48(2), pp. 170–77.

Miller, J. D., J. McCain, D. R. Lynam, L. R. Few, B. Gentile, J. MacKillop, and W. K. Campbell. A comparison of the criterion validity of popular measures of narcissism and narcissistic personality disorder via the use of expert ratings. *Psychological Assessment*, 2014, vol. 26(3), pp. 958–69.

Pailing, A., J. Boon, and V. Egan. Personality, the Dark Triad, and violence. *Personality and Individual Differences*, 2014, vol. 67, pp. 81–86.

Reidy, D. E., A. Zeichner, J. D. Foster, and M. A. Martinez. Effects of narcissistic entitlement and exploitativeness on human physical aggression. *Personality and Individual Differences*, 2008, vol. 44(4), pp. 865–75.

Sedikides, C., E. A. Rudich, A. P. Gregg, M. Kumashiro, and C. Rusbult. Are normal narcissists psychologically healthy?: Self-esteem matters. *Journal of Personality and Social Psychology*, 2004, vol. 87(3), pp. 400–416.

Serin, R. C. Violent recidivism in criminal psychopaths. *Law and Human Behavior*, 1996, vol. 20(2), p. 207.

Stellwagen, K. K. Psychopathy, Narcissism, and Machiavellianism: Distinct yet intertwining personality constructs. In *Narcissism and Machiavellianism in Youth: Implications for the development of adaptive and maladaptive behavior*, C. T. Barry, P. K. Kerig, K. K. Stellwagen, T. D. Barry, editors, pp. 25–45. American Psychological Association, 2011.

Tolmacz, R., and M. Mikulincer. The sense of entitlement in romantic relationships—Scale construction, factor structure, construct validity, and its associations with attachment orientations. *Psychoanalytic Psychology*, 2011, vol. 28(1), p. 75.

Woodworth, M. and S. Porter. In cold blood: Characteristics of criminal homicides as a function of psychopathy. *Journal of Abnormal Psychology*, 2002, vol. 111(3), p. 436.

Zeigler-Hill, V., and A. Besser. A glimpse behind the mask: Facets of narcissism and feelings of self-worth. *Journal of Personality Assessment*, 2013, vol. 95(3), pp. 249–60.

Zeigler-Hill, V., B. Enjaian, and L. Essa. The role of narcissistic personality features in sexual aggression. *Journal of Social and Clinical Psychology*, 2013, vol. 32(2), pp. 186–99.

Zeigler-Hill, V., E. M. Myers, and C. B. Clark. Narcissism and self-esteem reactivity: The role of negative achievement events. *Journal of Research in Personality*, 2010, vol. 44(2), pp. 285–92.

7. Warning Signs: Staying Alert for Narcissists

Dufner, M., J. F. Rauthmann, A. Z. Czarna, and J. J. Denissen. Are narcissists sexy? Zeroing in on the effect of narcissism on short-term mate appeal. *Personality and Social Psychology Bulletin*, 2013, vol. 39(7), pp. 870–82.

Hepper, E. G., R. H. Gramzow, and C. Sedikides. Individual differences in self-enhancement and self-protection strategies: An integrative analysis. *Journal of Personality*, 2010, vol. 78(2), pp. 781–814.

Holtzman, N. S., and M. J. Strube. Narcissism and attractiveness. *Journal of Research in Personality*, 2010, vol. 44(1), pp. 133–36.

Holtzman, N. S., and M. J. Strube. People with dark personalities tend to create a physically attractive veneer. *Social Psychological and Personality Science*, 2013, vol. 4(4), pp. 461–67.

8. Change and Recovery: Dealing with Lovers, Family, and Friends

Baskin-Sommers, A., E. Krusemark, and E. Ronningstam. Empathy in narcissistic personality disorder: From clinical and empirical perspectives. *Personality Disorders: Theory Research, and Treatment*, 2014, vol. 5(3), pp. 323–33.

Finkel, E. J., W. K. Campbell, L. E. Buffardi, M. Kumashiro, and C. E. Rusbult. The metamorphosis of Narcissus: Communal activation promotes relationship commitment among narcissists. *Personality and Social Psychology Bulletin*, 2009, vol. 35(10), pp. 1271–84.

Foster, J. D., I. Shrira, and W. K. Campbell. Theoretical models of narcissism, sexuality, and relationship commitment. *Journal of Social and Personal Relationships*, 2006, vol. 23(3), pp. 367–86.

Giacomin, M., and C. H. Jordan. Down-regulating narcissistic tendencies: communal focus reduces state narcissism. *Personality and Social Psychology Bulletin*, 2014, vol. 40(4), pp. 488–500.

Hatfield, E., and R. L. Rapson. *Love, Sex, and Intimacy: Their psychology, biology, and history.* HarperCollins College Publishers, 1993.

Hepper, E. G., C. M. Hart, and C. Sedikides. Moving Narcissus: Can narcissists be empathic? *Personality and Social Psychology Bulletin*, 2014, vol. 40 (9), pp. 1079–91.

Johnson, S. M. *Hold Me Tight: Seven conversations for a lifetime of love.* Little, Brown, 2008.

Johnson, S. M. *Love Sense: The revolutionary new science of romantic relationships.* Little, Brown, 2013.

Keller, P. S., S. Blincoe, L. R. Gilbert, C. N. Dewall, E. A. Haak, and T. Widiger. Narcissism in romantic relationships: A dyadic perspective. *Journal of Social and Clinical Psychology*, 2014, vol. 33(1), pp. 25–50.

Konrath, S., B. J. Bushman, and W. K. Campbell. Attenuating the link between threatened egotism and aggression. *Psychological Science*, 2006, vol. 17(11), pp. 995–1001.

Murray, S. L., J. G. Holmes, and D. W. Griffin. The benefits of positive illusions: Idealization and the construction of satisfaction in close relationships. *Journal of Personality and Social Psychology*, 1996, vol. 70(1), p. 79.

Pincus, A. L., N. M. Cain, and A. G. Wright. Narcissistic grandiosity and narcissistic vulnerability in psychotherapy. *Personality Disorders: Theory, Research, and Treatment*, 2014, vol. 5(4), pp. 439–43.

Vazire, S., L. P. Naumann, P. J. Rentfrow, and S. D. Gosling. Portrait of a narcissist: Manifestations of narcissism in physical appearance. *Journal of Research in Personality*, 2008, vol. 42(6), pp. 1439–47.

9. Coping and Thriving:
Dealing with Colleagues and Bosses

Bartlett, J. E., and M. E. Bartlett. Workplace bullying: An integrative literature review. *Advances in Developing Human Resources*, 2011, vol. 13(1), pp. 69–84.

Campbell, W. K., B. J. Hoffman, S. M. Campbell, and G. Marchisio. Narcissism in organizational contexts. *Human Resource Management Review*, 2011, vol. 21(4), pp. 268–84.

DuBrin, A. J. *Narcissism in the Workplace: Research, opinion, and practice.* Edward Elgar, 2012.

Grijalva, E., and D. A. Newman. Narcissism and Counterproductive Work Behavior (CWB): Meta-analysis and consideration of collectivist culture, Big Five personality, and narcissism's facet structure. *Applied Psychology*, 2014, in press.

Harvey, P., and M. J. Martinko. An empirical examination of the role of attributions in psychological entitlement and its outcomes. *Journal of Organizational Behavior*, 2009, vol. 30(4), pp. 459–76.

Higgs, M. The good, the bad and the ugly: Leadership and narcissism. *Journal of Change Management*, 2009, vol. 9(2), pp. 165–78.

Konrath, S., B. J. Bushman, and W. K. Campbell. Attenuating the link between threatened egotism and aggression. *Psychological Science*, 2006, vol. 17(11), pp. 995–1001.

Namie, G., and R. Namie. *The Bully at Work: What you can do to stop the hurt and reclaim your dignity on the job.* Sourcebooks, 2009.

Nevicka, B., A. H. De Hoogh, A. E. Van Vianen, B. Beersma, and D. McIlwain. All I need is a stage to shine: Narcissists' leader emergence and performance. *The Leadership Quarterly*, 2011, vol. 22(5), pp. 910–25.

O'Boyle Jr., E. H., D. R. Forsyth, G. C. Banks, and M. A. McDaniel. A meta-analysis of the dark triad and work behavior: A social exchange perspective. *Journal of Applied Psychology*, 2012, vol. 97(3), p. 557.

Padilla, A., R. Hogan and R. B. Kaiser. The toxic triangle: Destructive leaders, susceptible followers, and conducive environments. *The Leadership Quarterly*, 2007, vol. 18(3), pp. 176–94.

Penney, L. M., and P. E. Spector. Narcissism and counterproductive work behavior: Do bigger egos mean bigger problems? *International Journal of Selection and Assessment*, 2002, vol. 10(1, 2), pp. 126–34.

Spain, S. M., P. Harms, and J. M. LeBreton. The dark side of personality at work. *Journal of Organizational Behavior*, 2014, vol. 35(S1), pp. S41-S60.

Wesner, B. S. *Responding to the Workplace Narcissist.* Indiana University Press, 2007.

10. Advice for Parents: Raising a Confident, Caring Child

Baumrind, D. Child-care practices anteceding three patterns of preschool behavior. *Genetic Psychology Monographs*, 1967 vol. 75, pp. 43–88.

Brown, K. M., R. Hoye, and M. Nicholson. Self-esteem, self-efficacy, and social connectedness as mediators of the relationship between volunteering and well-being. *Journal of Social Service Research*, 2012, vol. 38(4), pp. 468–83.

Cramer, P. Young. Adult narcissism: A 20 year longitudinal study of the contribution of parenting styles, preschool precursors of narcissism, and denial. *Journal of Research in Personality*, 2011, vol. 45(1), pp. 19–28. The four parenting style descriptors in this chapter are taken in part or adapted from Cramer's analysis.

Choi, Y., Y. S. Kim, S. Y. Kim, and I. K. Park. Is Asian American parenting controlling and Harsh? Empirical testing of relationships between Korean American and Western parenting measures. *Asian American Journal of Psychology*, 2013, vol. 4(1), pp. 19–29.

Deater-Deckard, K. Tiger parents, other parents. *Asian American Journal of Psychology*, 2013, vol. 4(1), pp. 76–78.

Gavazzi, I. G., and V. Ornaghi. Emotional state talk and emotion understanding: A training study with preschool children. *Journal of Child Language*, 2011, vol. 38(5), pp. 1124–39.

Gelb, C. M. *The Relationship Between Empathy and Attachment in the Adolescent Population*. Dissertation Abstracts International: Section B: The Sciences and Engineering, 2002, vol. 62(9-B), p. 4252.

Goldstein, T. R., and E. Winner. Enhancing empathy and theory of mind. *Journal of Cognition and Development*, 2012, vol. 13(1), pp. 19–37.

Henry, C. S., and L. Hubbs-Tait. New directions in authoritative parenting. In *Authoritative Parenting: Synthesizing nurturance and discipline for optimal child development*, R. E. Larzelere, A. S. Morris, A. W. Harrist, editors, pp. 237–64. American Psychological Association, 2013.

Horton, R. S., and T. Tritch. Clarifying the links between grandiose narcissism and parenting. *Journal of Psychology*, 2014, vol. 148(2), pp. 133–43.

Juang, L. P., D. B. Qin, and I. J. Park. Deconstructing the myth of the "tiger mother": An introduction to the special issue on tiger parenting, Asian-heritage families, and child/adolescent well-being. *Asian American Journal of Psychology*, 2013, vol. 4(1), p 1.

Kidd, D. C., and E. Castano. Reading literary fiction improves theory of mind. *Science*, 2013, vol. 342(6156), pp. 377–80.

Kim, S. Y., Y. Wang, D. Orozco-Lapray, Y. Shen, and M. Murtuza. Does "Tiger Parenting" Exist? Parenting profiles of Chinese Americans and adolescent developmental outcomes. *Asian American Journal of Psychology*, 2013, vol. 4(1), pp. 7–18.

Maccoby, E. E., and J. A. Martin. Socialization in the context of the family: Parent–child interaction. *Handbook of Child Psychology: Vol. 4. Socialization, personality, and social development* (4th ed.), P. H. Mussen and E. M. Hetherington, editors. Wiley, 1983.

Mechanic, K. L., and C. T. Barry. Adolescent grandiose and vulnerable narcissism: Associations with perceived parenting practices. *Journal of Child and Family Studies*, April 2014, pp. 1–9.

Mongrain, M., J. M. Chin, and L. B. Shapira. Practicing compassion increases happiness and self-esteem. *Journal of Happiness Studies*, 2011, vol. 12(6), pp. 963–81.

Omer, H., S. G. Steinmetz, T. Carthy, and A. von Schlippe. The anchoring function: Parental authority and the parent-child bond. *Family Process*, 2013, vol. 52(2), pp. 193–206.

Ornaghi, V., J. Brockmeier, and I. Grazzani. Enhancing social cognition by training children in emotion understanding: a primary school study. *Journal of Experimental Child Psychology*, 2014, vol. 119, pp. 26–39.

Padilla-Walker, L. M., and L. J. Nelson. Black Hawk down? Establishing helicopter parenting as a distinct construct from other forms of parental control during emerging adulthood. *Journal of Adolescence*, 2012, vol. 35(5), pp. 1177–90.

Phelan, T. *1-2-3 Magic: Effective discipline for children 2–12*. ParentMagic, Inc., 2010.

Ryder, J. A. *College Student Volunteerism: A quantitative analysis of psychological benefits gained through time spent in service to others*. ProQuest Information & Learning, 2006, dissertation 66.

Schiffrin, H. H., M. Liss, H. Miles-McLean, K. A. Geary, M. J. Erchull, and T. Tashner. Helping or hovering? The effects of helicopter parenting on college students' well-being. *Journal of Child and Family Studies*, 2014, vol. 23(3), pp. 548–57.

Segrin, C., A. Woszidlo, M. Givertz, and N. Montgomery. Parent and child traits associated with overparenting. *Journal of Social and Clinical Psychology*, 2013, vol. 32(6), pp. 569–95.

Supple, A. J., and A. M. Cavanaugh. Tiger mothering and Hmong American parent-adolescent relationships. *Asian American Journal of Psychology*, 2013, vol. 4(1), p. 41.

Vinik, J., A. Almas, and J. Grusec. Mothers' knowledge of what distresses and what comforts their children predicts children's coping, empathy, and prosocial behavior. *Parenting: Science and Practice*, 2011, vol. 11(1), pp. 56–71.

11. SoWe: The Healthy Use of Social Media

Ahn, D., and D.H. Shin. Is the social use of media for seeking connectedness or for avoiding social isolation? Mechanisms underlying media use and subjective well-being. *Computers in Human Behavior*, 2013, vol. 29(6), pp. 2453–62.

Arthur, C. A., and L. Hardy. Transformational leadership: A quasi-experimental study. *Leadership & Organization Development Journal*, 2014, vol. 35(1), pp. 38–53.

Bergman, S. M., M. E. Fearrington, S. W. Davenport, and J. Z. Bergman. Millennials, narcissism, and social networking: What narcissists do on social networking sites and why. *Personality and Individual Differences*, 2011, vol. 50(5), pp. 706–11.

Buckels, E. E., P. D. Trapnell, and D. L. Paulhus. Trolls just want to have fun. *Personality and Individual Differences*, 2014, vol. 67, pp. 97–102.

Buffardi, L. E., and W. K. Campbell. Narcissism and social networking web sites. *Personality and Social Psychology Bulletin*, 2008, vol. 34(10), pp. 1303–14.

Burke, M., C. Marlow, and T. Lento. *Social Network Activity and Social Well-Being*. Proceedings of the SIGCHI Conference on Human Factors in Computing Systems, ACM, 2010.

Burlingame, G. M., D. T. McClendon, and J. Alonso. Cohesion in group therapy. *Psychotherapy*, 2011, vol. 48(1), pp. 34–42.

Carpenter, C. J. Narcissism on Facebook: Self-promotional and anti-social behavior. *Personality and Individual Differences*, 2012, vol. 52(4), pp. 482–86.

Chou, H. T., and N. Edge. They are happier and having better lives than I am": The impact of using Facebook on perceptions of others' lives. *Cyberpsychology, Behavior, and Social Networking*, 2012, vol. 15(2), pp. 117–21.

Ellison, N. B., C. Steinfield, and C. Lampe. The benefits of Facebook "friends": Social capital and college students' use of online social network sites. *Journal of Computer-Mediated Communication*, 2007, vol. 12(4), pp. 1143–68.

Gentile, B., J. M. Twenge, E. C. Freeman, and W. K. Campbell. The effect of social networking websites on positive self-views: An experimental investigation. *Computers in Human Behavior*, 2012, vol. 28(5), pp. 1929–33.

Gonzales, A. L., and J. T. Hancock. Mirror, mirror on my Facebook wall: Effects of exposure to Facebook on self-esteem. *Cyberpsychology, Behavior, and Social Networking*, 2011, vol. 14(1, 2), pp. 79–83.

Greenwood, D. N. Fame, Facebook, and Twitter: How attitudes about fame predict frequency and nature of social media use. *Psychology of Popular Media Culture*, 2013, vol. 2(4), p. 222.

Hanckel, B., and A. Morris. Finding community and contesting heteronormativity: Queer young people's engagement in an Australian online community. *Journal of Youth Studies*, 2014, vol. 17(7), pp. 872–86.

Kross, E., P. Verduyn, E. Demiralp, J. Park, D. S. Lee, N. Lin, H. Shablack, J. Jonides, and O. Ybarra. Facebook use predicts declines in subjective well-being in young adults. *PLoS One*, 2013, vol. 8(8), e69841.

Malkin, C., and J. E. Stake. Changes in attitudes and self-confidence in the women's and gender studies classroom: The role of teacher alliance and student cohesion. *Sex Roles*, 2004, vol. 50(7, 8), pp. 455–68.

Manago, A. M., T. Taylor, and P. M. Greenfield. Me and my 400 friends: The anatomy of college students' Facebook networks, their communication patterns, and well-being. *Developmental Psychology*, 2012, vol. 48(2), pp. 369–80.

Nathan DeWall, C., L. E. Buffardi, I. Bonser, and W. Keith Campbell. Narcissism and implicit attention seeking: Evidence from linguistic analyses of social networking and online presentation. *Personality and Individual Differences*, 2011, vol. 51(1), pp. 57–62.

Panek, E. T., Y. Nardis, and S. Konrath. Mirror or megaphone?: How relationships between narcissism and social networking site use differ on Facebook and Twitter. *Computers in Human Behavior*, 2013, vol. 29(5), pp. 2004–12.

Song, H., A. Zmyslinski-Seelig, J. Kim, A. Drent, A. Victor, K. Omori, and M. Allen. Does Facebook make you lonely?: A meta analysis. *Computers in Human Behavior*, 2014, vol. 36, pp. 446–52.

Weiser, E. B. The functions of Internet use and their social and psychological consequences. *CyberPsychology & Behavior*, 2001, vol. 4(6), pp. 723–43.

Wilson, R. E., S. D. Gosling, and L. T. Graham. A review of Facebook research in the social sciences. *Perspectives on Psychological Science*, 2012, vol. 7(3), pp. 203–20.

12. A Passionate Life: The Ultimate Gift of Healthy Narcissism

Frimer, J. A., L. J. Walker, W. L. Dunlop, B. H. Lee, and A. Riches. The integration of agency and communion in moral personality: Evidence of enlightened self-interest. *Journal of Personality and Social Psychology*, 2011, vol. 101(1), pp. 149–63.

Greenberg, J., and Mitchell, S. *Object Relations in Psychoanalytic Theory*. Harvard University Press, 1983.

INDEX

ABOUT THE AUTHOR

DR. CRAIG MALKIN is an author, clinical psychologist, and lecturer for Harvard Medical School with two decades of experience helping individuals, couples, and families. His articles, advice, and insights on relationships have appeared in newspapers and magazines such as *Time*, the *New York Times*, *The Sunday Times* (London), *Psychology Today*, and *Women's Health*, as well as countless other popular print and online media outlets. He's also been featured multiple times on NPR, CBS Radio, and the Oprah Winfrey Network channel; appeared on various local morning shows; and been a guest on more than sixty radio stations in the United States and abroad. Dr. Malkin is president and director of the Cambridge, Massachusetts–based YM Psychotherapy and Consultation, Inc., which provides psychotherapy and evidence-based couples workshops. He lives in Boston with his wife and twin girls.